Atlas of Epilepsy Surgery

Takehiro Uda • Shunsuke Nakae
Daijiro Kojima

Atlas of Epilepsy Surgery

Anatomical Understanding with Illustrations

Springer

Takehiro Uda
Graduate School of Medicine
Department of Neurosurgery
Osaka Metropolitan University
Abeno-ku, Osaka, Osaka, Japan

Shunsuke Nakae
Department of Neurosurgery
Fujita Medical University School
of Medicine
Toyoake, Aichi, Aichi, Japan

Daijiro Kojima
Department of Neurosurgery
Fujita Medical University School
of Medicine
Toyoake, Aichi, Aichi, Japan

ISBN 978-981-92-0175-4 ISBN 978-981-92-0176-1 (eBook)
https://doi.org/10.1007/978-981-92-0176-1

This Springer imprint is published by the registered company Springer Nature Singapore Pte Ltd.
The registered company address is: 152 Beach Road, #21-01/04 Gateway East, Singapore 189721, Singapore

Foreword by Kato

This book systematically compiles contemporary knowledge and practical techniques in epilepsy surgery, grounded in extensive clinical experience.

The authors possess broad expertise in epilepsy surgery not only in Japan but also internationally. Drawing on perspectives cultivated at the forefront of the field, they provide a clear and thoughtful explanation of treatment strategies and surgical techniques from both theoretical and practical viewpoints. A distinctive feature of this book is that it extends beyond a mere compilation of surgical procedures, clearly elucidating the fundamental principles that guide surgical decision-making, grounded in anatomical and functional considerations.

In particular, the emphasis on minimally invasive approaches and functional preservation, together with up-to-date treatment strategies, offers valuable insights into the future direction of epilepsy surgery.

This book will serve as an indispensable guide not only for young neurosurgeons embarking on a career in epilepsy surgery but also for experienced specialists seeking to reassess and further refine their clinical practice.

It is highly recommended for neurosurgeons, epileptologists, and other specialists involved in epilepsy surgery and functional neurosurgery.

Yoko Kato

President, International Society of Minimally
Invasive Neurosurgery (ISMINS)
Tashkent, Uzbekistan

President, Asian Congress of Neurological Surgeons
Kaohsiung, Taiwan

Assistant Director, Fujita Health University (FHU)
Bantane Hospital
Nagoya, Aichi, Japan

Professor and Chair, Department of Neurosurgery
FHU Bantane Hospital
Nagoya, Aichi, Japan

Head of FHU Bantane Hospital, Stroke Center
Nagoya, Aichi, Japan

Foreword by Morino

In epilepsy surgery, precise anatomical knowledge of the brain, cranial nerves, and vascular structures is indispensable. Epilepsy surgery is fundamentally functional in nature; therefore, every effort must be made to minimize surgical complications, and meticulous operative technique is required.

Dr. Uda is a junior alumnus of my alma mater, Osaka City University (now Osaka Metropolitan University), and trained in epilepsy surgery under my supervision during my tenure as Chief of the Department at Tokyo Metropolitan Neurological Hospital. Dr. Nakae of Fujita Health University likewise received training in epilepsy surgery during my time at Kumagaya General Hospital. These two physicians are now leading epilepsy surgery teams at their respective institutions. Dr. Kojima of Fujita Health University is an accomplished neurosurgeon and an outstanding neuroillustrator who created the illustrations that form the core of this book.

This book covers representative epilepsy surgical procedures, with exceptionally clear and detailed illustrations of the relevant surgical anatomy at each step of the operation. It provides the essential knowledge required for safe and effective surgery and will serve as an invaluable reference for physicians beginning their training in epilepsy surgery. With confidence in its scientific rigor and educational value, and with hope for the continued advancement of epilepsy surgery, I am pleased to recommend this book.

Michiharu Morino

Director, Morino Clinic
Minato-ku, Tokyo, Japan

Former Associate Professor
Department of Neurosurgery
Osaka Metropolitan University
Osaka, Japan

Former Chief, Department of Neurosurgery
Tokyo Metropolitan Neurological Hospital
Fuchu, Japan

Former Honorary Director
Department of Neurosurgery
Kumagaya General Hospital
Kumagaya, Japan

Preface: Introduction of Atlas of Epilepsy Surgery

Epilepsy surgery can be broadly classified into three main categories: resective surgery, which removes the epileptogenic focus; disconnection surgery, which interrupts the propagation of epileptic discharges; and palliative surgery, which aims to reduce seizure frequency or severity without complete removal of the epileptogenic zone. Accurate localization of the epileptogenic zone is essential for selecting the appropriate surgical strategy. When noninvasive diagnostic methods are insufficient to localize the epileptogenic area, invasive monitoring with intracranial electrodes is employed.

In recent years, minimally invasive approaches such as laser interstitial thermal therapy (LITT) and radiofrequency thermocoagulation (RFTC), as well as neuromodulation therapies including deep brain stimulation (DBS) and responsive neurostimulation (RNS), have gained increasing attention. Nevertheless, from the perspectives of efficacy, immediacy, and cost-effectiveness, open epilepsy surgery continues to play a central role.

As a form of functional neurosurgery, epilepsy surgery must be performed with safety, precision, and functional preservation. Because epilepsy surgery often relies on functional modification rather than extensive anatomical resection, and the procedures are relatively standardized, mastering epilepsy surgery requires a three-dimensional understanding of neuroanatomy. To achieve this, off-the-job training methods such as reviewing surgical videos, performing cadaveric dissections, and studying surgical atlases are indispensable. Among these, surgical atlases are particularly valuable because they allow learners to visualize critical steps from a three-dimensional perspective and review them repeatedly in a time-efficient manner.

This textbook was written by three experienced epilepsy surgeons with the goal of providing a practical and comprehensive resource for learning epilepsy surgery. We hope that it will contribute to further improving surgical safety, precision, and minimally invasive techniques, ultimately leading to better seizure control for patients with epilepsy.

Finally, we would like to express our sincere gratitude to the two distinguished mentors who kindly provided recommendation letters for this book. Professor Yoko Kato gave us invaluable opportunities to perform epilepsy surgery overseas and inspired us to write this textbook in English. Dr. Michiharu Morino, our mentor, taught us the principles and techniques of epilepsy surgery and has profoundly influenced our surgical practice. We also extend our heartfelt thanks to Noeru Kawase of the Fujita Health University School of Medicine, who primarily contributed to the English editing of this textbook.

February 2026

Takehiro Uda
Shunsuke Nakae
Daijiro Kojima

Contents

About the Authors

Takehiro Uda, MD, PhD, serves as a neurosurgeon at Osaka Metropolitan University, specializing in functional neurosurgery. He delivers comprehensive care to both pediatric and adult patients and possesses extensive experience across the full spectrum of epilepsy surgery. His clinical philosophy centers on achieving safe and reliable surgical outcomes while preserving neurological function. He is widely recognized as a leading authority in epilepsy surgery and contributes significantly to research, education, and the advancement of modern neurosurgical practice.

Shunsuke Nakae, MD, MSc, PhD, is an Associate Professor in the Department of Neurosurgery at Fujita Health University, Japan. He specializes in epilepsy and glioma surgery, with particular expertise in stereoelectroencephalography (SEEG), awake craniotomy, and minimally invasive techniques.

He received advanced training in SEEG at the Claudio Munari Centre for Epilepsy Surgery, Niguarda Hospital in Milan, Italy, and is actively engaged in international collaborations across Asia. His research focuses on epilepsy and glioma, including functional brain mapping, neuroimaging, genetic analysis, and the development of safer and more precise surgical strategies.

Dr. Nakae is also involved in educational and clinical initiatives to expand access to epilepsy surgery.

Daijiro Kojima, is a neurosurgeon in the Department of Fujita Health University School of Medicine, Japan. He specializes in epilepsy surgery, surgery for malignant brain tumors, and awake surgery, and has extensive experience performing these procedures.

Dr. Kojima is skilled in creating illustrations and was responsible for producing the figures in this book. In creating the figures for this book, he aimed to produce illustrations that combine anatomical clarity with the sense of realism that only a surgeon can convey.

Dr. Kojima has contributed numerous figures to peer-reviewed papers and conference presentations. He is well-versed in higher brain functions and the pathways of nerve fibers within the brain, and through his expertise and illustrations, he is working to facilitate safe preoperative simulations.

1 Anterior Temporal Lobectomy (ATL)

1.1 Surgical Procedures

1.1.1 Patient Positioning and Head Fixation

The patient is placed in the supine position, with the head rotated 15–20 degrees to the contralateral side and tilted downward (vertex down) for fixation (Figs. 1.1 and 1.2). Slight head extension aligns the visual axis with the long axis of the hippocampus (Fig. 1.3).

1.1.2 Skin Incision and Craniotomy

A curvilinear skin incision is made from the preauricular area to the midline, followed by a frontotemporal craniotomy (Fig. 1.4). The sphenoid ridge is flattened using a high-speed drill.

Fig. 1.1 Patient positioning for ATL. The patient is placed in the supine position, with the head rotated 15–20 degrees to the contralateral side

T. Uda et al., *Atlas of Epilepsy Surgery*, https://doi.org/10.1007/978-981-92-0176-1_1

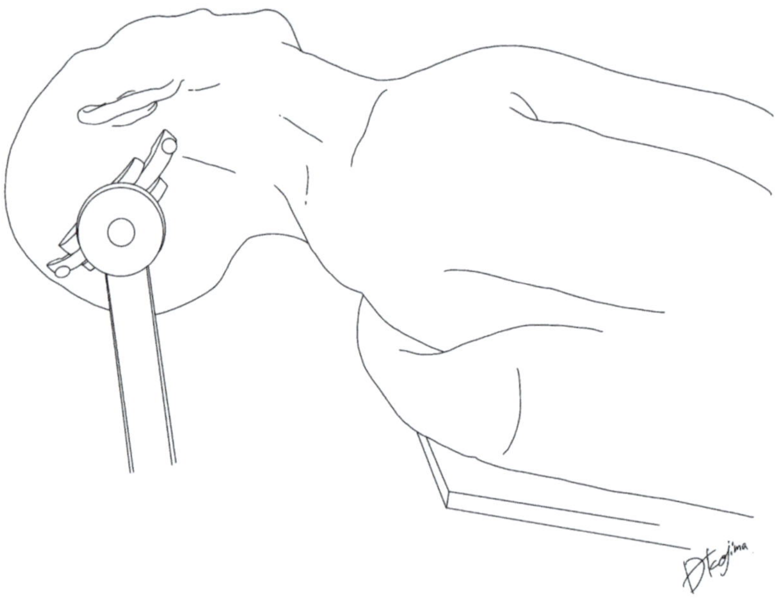

Fig. 1.2 Patient positioning for ATL. The head is fixed in a vertex-down position

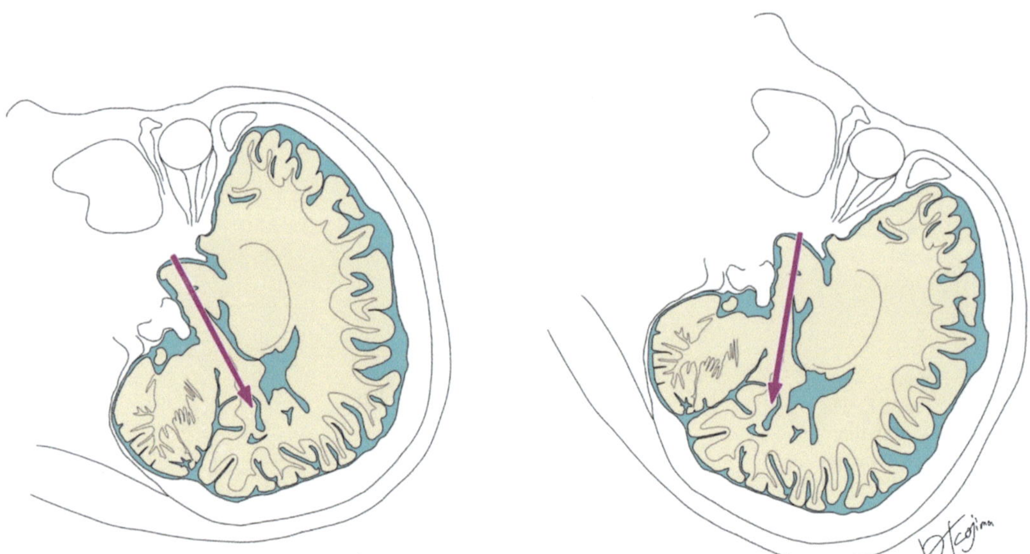

Fig. 1.3 Rationale for vertex-down positioning in ATL. Horizontal head positioning (left), the usual microscopic trajectory from the vertex, results in misalignment between the visual axis and the long axis of the hippocampus. With vertex-down positioning (right), alignment of the visual axis with the hippocampal long axis can be achieved

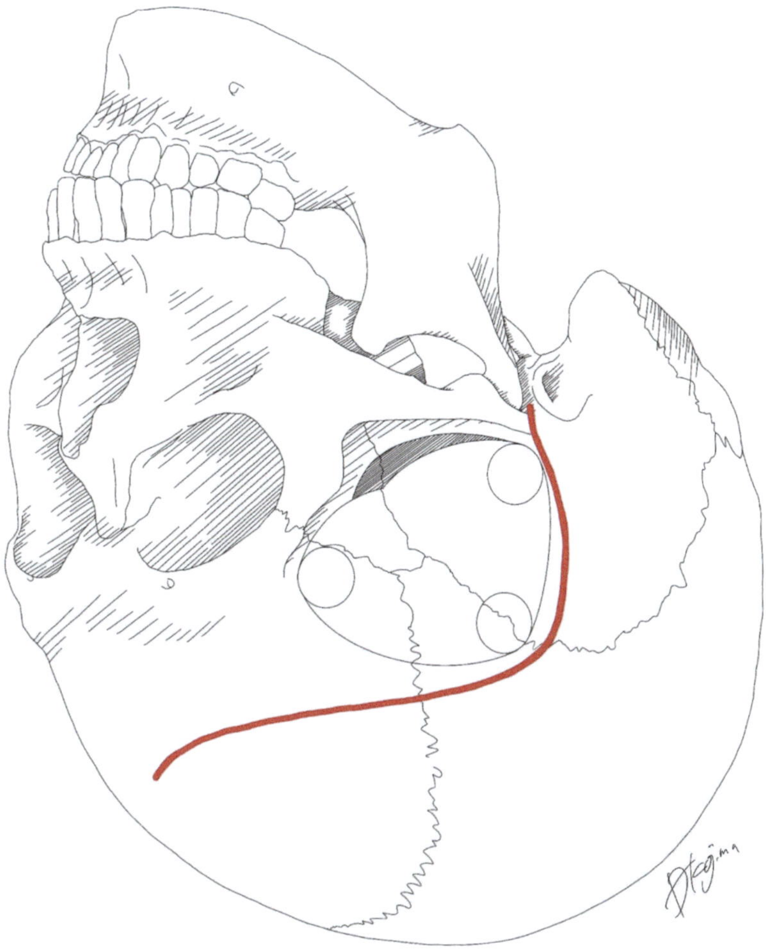

Fig. 1.4 Skin incision line and the extent of craniotomy for ATL. A curvilinear skin incision is made anterior to the auricle. Posterior extension in a question mark-shaped fashion is not required. Burr holes are placed at the frontozygomatic suture and at the root of the zygoma. An additional burr hole is created directly superior to the zygomatic burr hole on the linea temporalis to complete a frontotemporal craniotomy. The sphenoid ridge is drilled until the meningo-orbital band is identified

1.1.3 Resection of the Lateral Temporal Lobe and Amygdala

The procedure begins with resection of the anterior portion of the temporal lobe. In the dominant hemisphere, resection is limited to 35 mm from the temporal tip, whereas in the nondominant hemisphere, it can be extended up to 50 mm (Fig. 1.5). A cortical incision is made along the superior or middle temporal gyrus toward the Sylvian fissure (Fig. 1.6), and subpial resection of the temporal operculum is performed. Anteriorly, the operculum is resected to the middle fossa, and posteriorly, to the planned posterior margin of the lateral temporal resection (Fig. 1.7).

Through a subpial approach, the insula, the inferior trunk of the middle cerebral artery (MCA), and the inferior peri-insular sulcus are identified. Using the inferior peri-insular sulcus as a landmark, deep white matter dissection is carried out through the temporal stem until the inferior horn of the lateral ventricle is reached (Fig. 1.8).

Once the inferior horn is exposed, the incision in the temporal stem is extended both anteriorly and posteriorly. Anteriorly, it is extended to connect with the middle fossa; posteriorly, it is extended to reach the planned posterior margin of the lateral temporal resection. At this point, the amygdala should be removed, exposing the head and body of the hippocampus (Fig. 1.9).

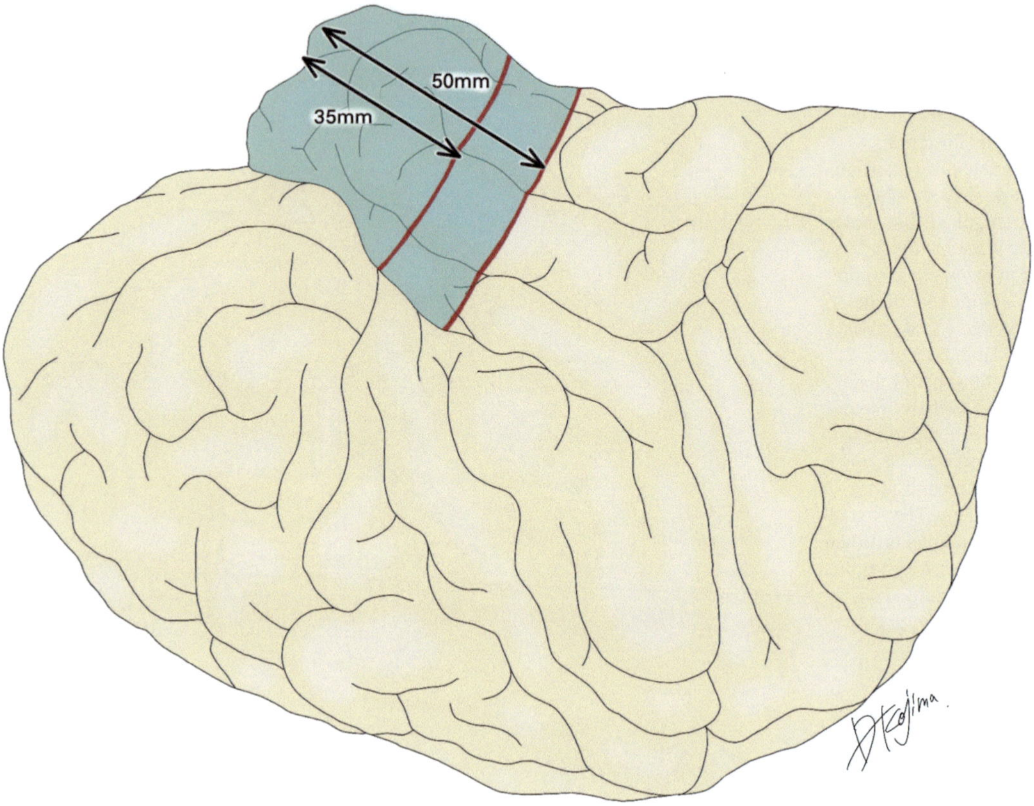

Fig. 1.5 Extent of temporal pole resection in the dominant and nondominant hemispheres. Although individual variation exists, resection of the temporal pole is generally limited to approximately 35 mm from the temporal tip in the dominant hemisphere, whereas it may be extended up to 50 mm in the nondominant hemisphere

1.1.4 Resection of the Hippocampus

The uncus is dissected subpially from the ventricular side toward the cisternal side. At this stage, important landmarks that can be identified through the pia mater include the optic tract, oculomotor nerve, and anterior choroidal artery. The medial aspect of the hippocampal head is transected toward the inferior choroidal point of the choroidal fissure, allowing the removal of the uncus.

Next, dissection proceeds posteriorly. The fimbria is bluntly dissected with a dissector, and the lateral margin of the hippocampus is transected along the innominate sulcus toward the posterior end. Finally, the hippocampal tail is resected in a medial-to-lateral direction, and the hippocampus is removed en bloc as far as possible. The posterior limit of resection is generally defined at the level of the quadrigeminal cistern.

1.1.5 Closure

After confirming hemostasis, the dura is closed with sutures, and the bone flap is replaced to complete the procedure.

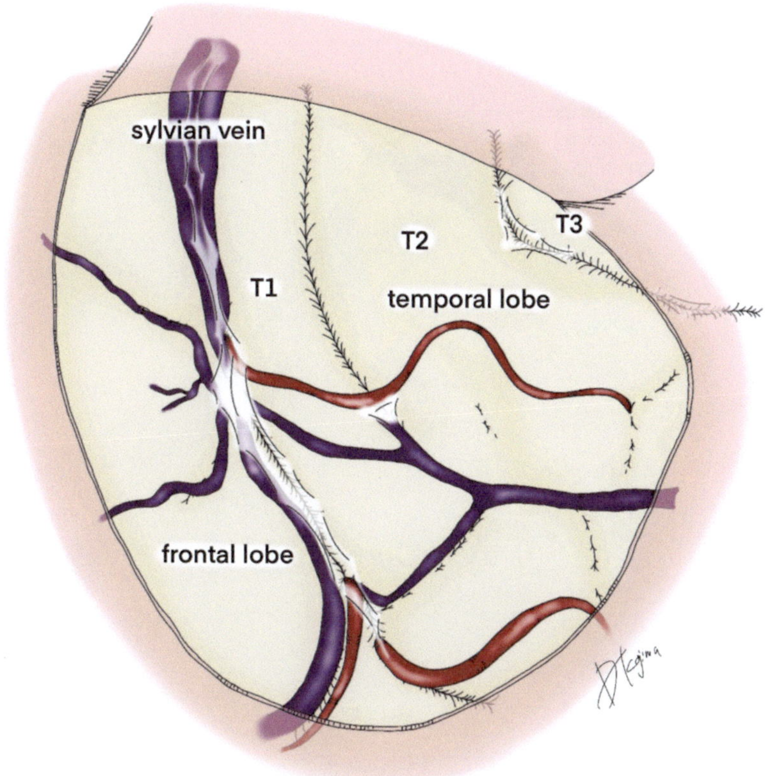

Fig. 1.6 Exposure of the cortical surface of the frontal and temporal lobes following a frontotemporal craniotomy. A cortical incision is made along the superior (T1) or middle temporal gyrus (T2) toward the Sylvian fissure, and resection of the temporal operculum is then initiated

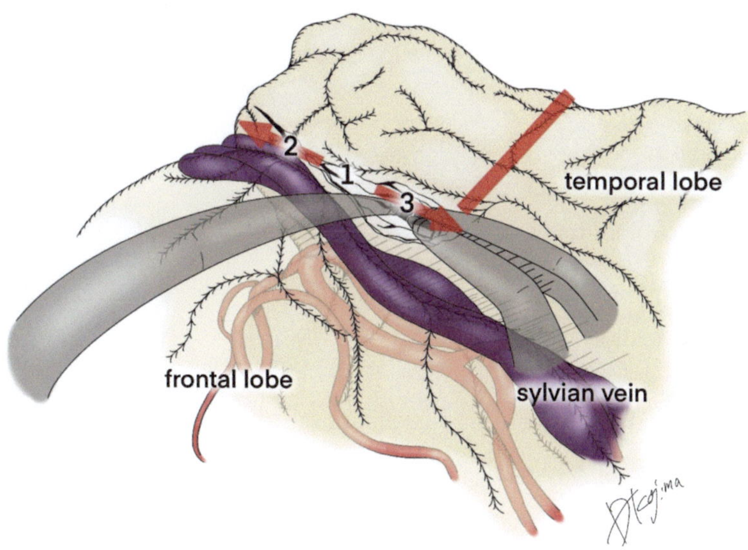

Fig. 1.7 Stepwise resection of the temporal tip. Following subpial resection, the operculum is resected anteriorly down to the middle cranial fossa and posteriorly to the predetermined margin of the lateral temporal resection

Fig. 1.8 Approach to the inferior horn through the temporal stem. Deep white matter dissection is performed through the temporal stem to reach the inferior horn of the lateral ventricle

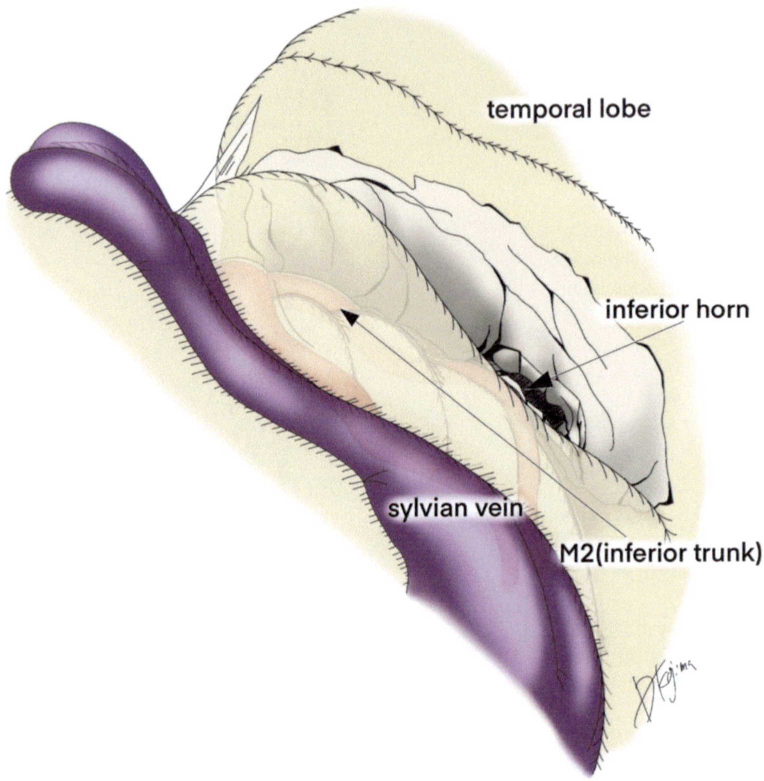

Fig. 1.9 Removal of the amygdala and exposure of the hippocampal head and body. After exposure of the inferior horn, the temporal stem incision is extended anteriorly to the middle fossa and posteriorly to the predetermined margin of the lateral temporal resection. Partial removal of the amygdala exposes the head and body of the hippocampus

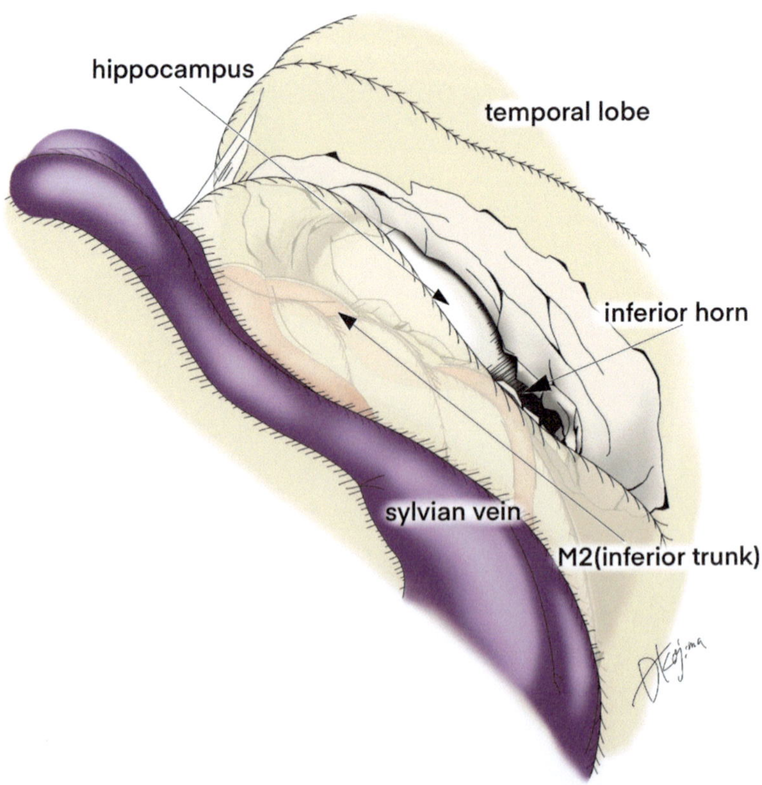

1.2 Notes

1.2.1 Anatomical Overview for ATL

In anterior temporal lobectomy (ATL), the lateral temporal cortex is resected to the level of the fusiform gyrus to secure an adequate working space for access to the medial temporal structures (Fig. 1.10). After dissection of the innominate sulcus toward the tentorium, the disconnection is advanced from the uncus to the inferior choroidal point, completing the medial disconnection of the hippocampal head (Fig. 1.11). The disconnection from the uncus to the inferior choroidal point is a crucial step common to the various surgical techniques described in this textbook (Fig. 1.12).

During hippocampectomy, vessels such as the middle hippocampal artery and posterior hippocampal artery are divided to remove the hippocampus.

After resection, key anatomical structures—including the optic nerve, optic tract, oculomotor nerve, posterior cerebral artery (PCA), and basal vein—can be visualized within the surgical cavity (Fig. 1.13).

1.2.2 Extent of Posterior Resection in the Superior Temporal Gyrus

The posterior limit of resection in the superior temporal gyrus differs between the dominant and nondominant hemispheres. In general, resection is limited to 35 mm from the temporal pole in the dominant hemisphere and 50 mm in the nondominant hemisphere. In the dominant hemisphere, intraoperative cortical mapping during awake craniotomy provides the most reliable method for determining the resection margin. In contrast, in pediatric or developmentally delayed patients for whom awake craniotomy is not feasible, cortico-cortical evoked potentials (CCEPs) can be a valuable adjunct to assess cortico-cortical connectivity.

In addition, the Medtronic S8 navigation system allows for the import of diffusion tensor imaging (DTI), enabling clear visualization of white matter tracts. This approach is particularly useful in cases where mass effect from a tumor causes displacement of fiber pathways.

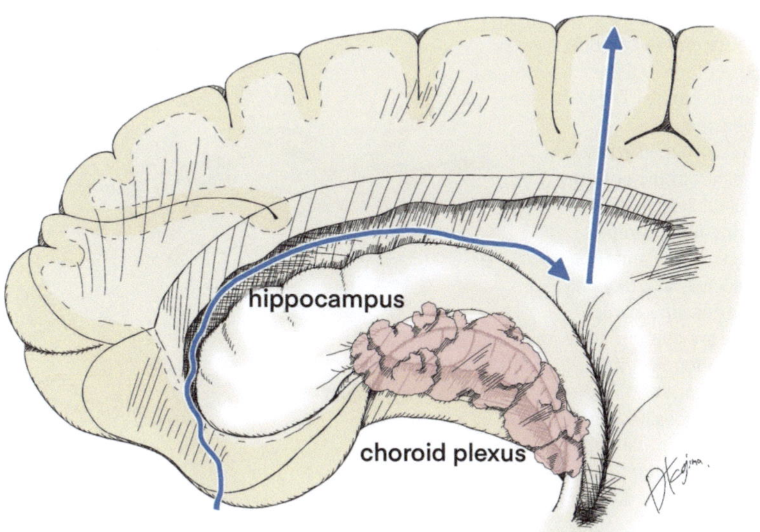

Fig. 1.10 Lateral temporal cortical resection in ATL. In ATL, the lateral temporal cortex is resected to the level of the fusiform gyrus to create sufficient working space for access to the medial temporal structures

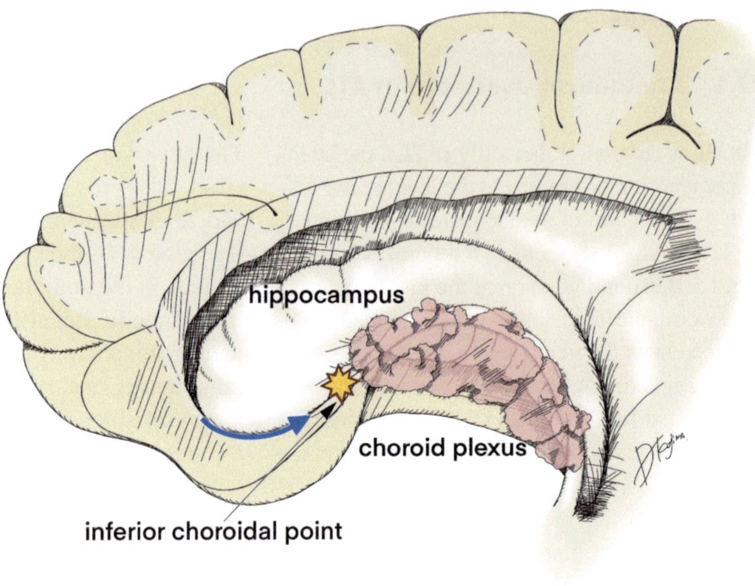

Fig. 1.11 Uncus-to-inferior choroidal point disconnection in ATL. Following lateral temporal cortical resection, the dissection proceeds from the uncus toward the inferior choroidal point, thereby completing the medial disconnection of the hippocampal head

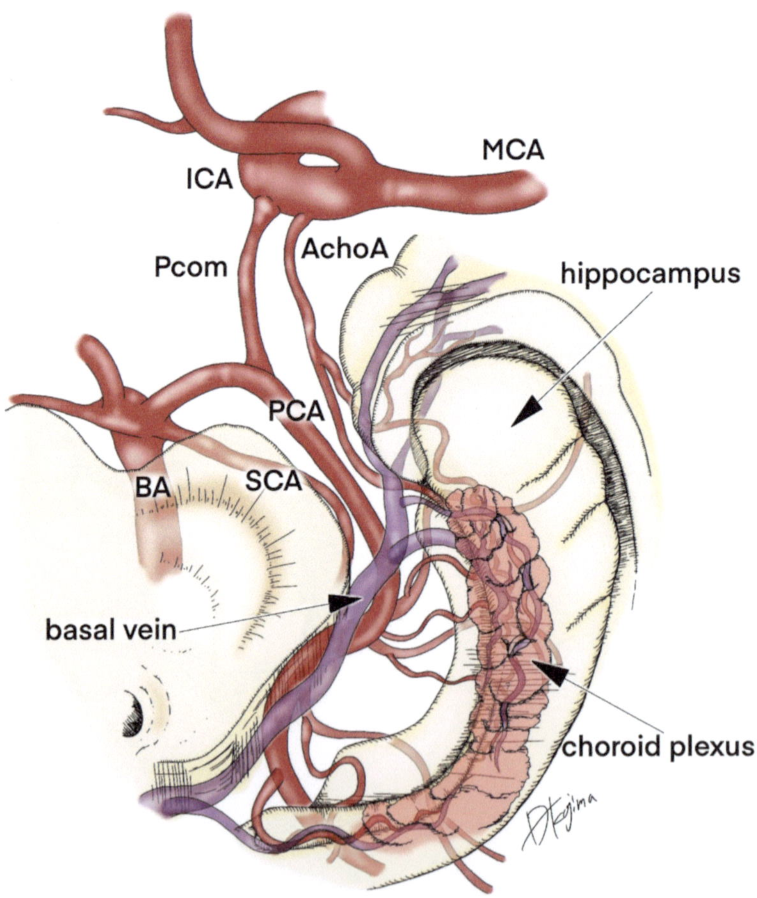

Fig. 1.12 Vascular anatomy of the medial temporal structures. This figure illustrates the vascular anatomy of the medial temporal structures. The anterior choroidal artery (AchoA) enters the inferior horn at the inferior choroidal point and continues as the choroid plexus. The hippocampus is primarily supplied by branches of the posterior cerebral artery (PCA). During hippocampal resection, the middle hippocampal artery and posterior hippocampal artery are appropriately divided to allow the removal of the hippocampus. (*AhoA* anterior choroidal artery, *BA* basilar artery, *ICA* internal carotid artery, *PCA* posterior cerebral artery, *Pcom* posterior communicating artery, *SCA* superior cerebellar artery)

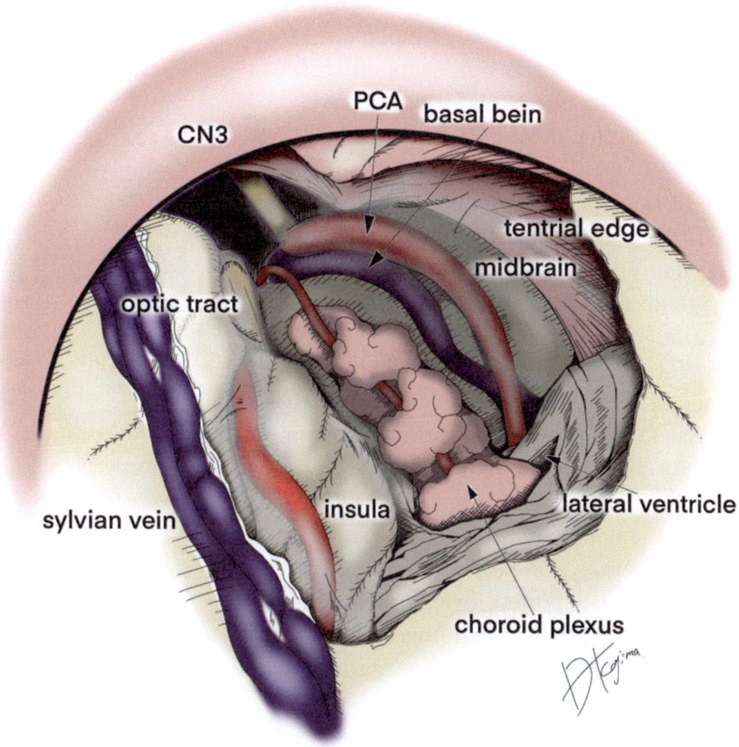

Fig. 1.13 Surgical view after resection in ATL. In the postresection surgical field, structures such as the optic nerve, optic tract, oculomotor nerve, posterior cerebral artery (PCA), and basal vein can be identified within the resection cavity (*CN2* optic nerve, *CN3* oculomotor nerve)

1.2.3 Sylvian Fissure Dissection

Dissection of the Sylvian fissure can lead to loss of mechanical support for the superior temporal gyrus, making the procedure more difficult. In addition, dissection of the Sylvian fissure inevitably prolongs operative time, which can be a disadvantage. On the other hand, in tumors with a strong mass effect involving the insula, the Sylvian fissure may appear relatively shallow, and there is a risk of unexpectedly encountering MCA branches at an unexpectedly shallow depth. In such cases, opening the Sylvian fissure allows direct visualization of the insular surface and confirmation of the course of these critical vessels, which can be advantageous. For these reasons, the decision to dissect the Sylvian fissure is at the surgeon's discretion according to each individual case.

1.2.4 Comparison with Selective Amygdalohippocampectomy (SAH)

Hu et al. compared outcomes between ATL and SAH in a meta-analysis of 13 comparative studies (total n = 1511). SAH was associated with lower odds of achieving seizure freedom than ATL (Engel class I: 66% vs 71%; odds ratio (OR) 0.65), while change in intelligence quotient at ~12 months showed no significant difference between groups [1].

Josephson et al. performed a meta-analysis of 13 comparative studies (total n = 1203) and also found that ATL achieved a higher rate of Engel class I outcomes than SAH (71% vs 63%; risk ratio (RR) 1.32), supporting the superiority of ATL in seizure control [2].

By contrast, Rangwala et al. analyzed up to 29 comparative studies (maximum total $n = 3238$) and found no significant difference in Engel class I seizure freedom between ATL and SAH overall (RR 0.96; 95% confidence interval (CI) 0.89–1.03; $P = 0.26$). However, the risk of visual field deficit was significantly lower after SAH (RR 0.87; 95% CI 0.78–0.97; $P = 0.01$), particularly with the transsylvian approach. Subgroup analyses suggested worse seizure outcomes with transsylvian SAH compared with ATL, whereas transcortical SAH achieved outcomes comparable to ATL. In analyses restricted to hippocampal sclerosis, no significant difference between ATL and SAH was detected [3].

Long-term follow-up also suggests a likely advantage for ATL in sustained seizure control. Bujarski et al. reported, over a mean 10-year follow-up, a longer duration Engel class IA status in ATL, with more recurrences during medication taper in the SAH group [4]. Mohan et al. similarly found a higher seizure freedom rate at 10 years after ATL, reinforcing evidence of a lower long-term relapse risk compared with SAH [5].

References

1. Hu WH, Zhang C, Zhang K, Meng FG. Selective amygdalohippocampectomy versus anterior temporal lobectomy in the management of mesial temporal lobe epilepsy: a meta-analysis of comparative studies. J Neurosurg. 2013;119(5):1089–97. https://doi.org/10.3171/2013.6.JNS121015.
2. Josephson CB, Dykeman J, Fiest KM, Liu X, Sadler RM, Jette N, et al. Systematic review and meta-analysis of standard vs selective temporal lobe epilepsy surgery. Neurology. 2013;80(18):1669–76. https://doi.org/10.1212/WNL.0b013e3182904f82.
3. Rangwala BS, Rangwala HS, Shafique MA, Mustafa MS, Danish F, Kumar S, et al. An examination of seizure-free outcome and visual field deficits: anterior temporal lobectomy versus selective amygdalohippocampectomy for temporal lobe epilepsy—a systematic review and meta-analysis for comprehensive understanding. Acta Neurochir. 2024;166:487. https://doi.org/10.1007/s00701-024-06383-6.
4. Bujarski KA, Hirashima F, Roberts DW, Jobst BC. Long-term seizure outcomes of anterior temporal lobectomy versus selective amygdalohippocampectomy in mesial temporal lobe epilepsy. J Neurosurg. 2013;119(1):16–23. https://doi.org/10.3171/2013.3.JNS121396.
5. Mohan M, Keller S, Nicolson A, Biswas S, Smith D, Osman Farah J, et al. The long-term outcomes of resective surgery for temporal lobe epilepsy: a systematic review and meta-analysis. PLoS One. 2018;13(2):e0190639. https://doi.org/10.1371/journal.pone.0190639.

2 Selective Amygdalohippocampectomy (SAH)

2.1 Surgical Procedure

2.1.1 Positioning and Fixation

The patient is placed in the supine position. After slight vertex-down positioning, the head is rotated approximately 20° toward the contralateral side. The vertex-down position is selected for two main reasons: first, mild neck extension facilitates the alignment of the surgeon's visual axis with the hippocampus's longitudinal axis; second, dissection of the Sylvian fissure is technically easier in this position (Fig. 1.3).

2.1.2 Skin Incision and Craniotomy

A semi-curvilinear skin incision is made, extending from the preauricular region to the frontal area. The temporalis muscle is dissected using the two-layer technique, separating the superficial and deep fascial layers.

Burr holes are created at three sites: one at the base of the frontozygomatic suture, one at the temporal base, and one posterior to the temporal line. A frontotemporal craniotomy is then performed. After craniotomy, the sphenoid ridge is extensively drilled to improve the visualization of the proximal Sylvian fissure. Illustrations of patient positioning, head fixation, and craniotomy design are provided in the ATL section (Figs. 1.1, 1.2, and 1.4).

2.1.3 Access to the Temporal Horn of the Lateral Ventricle

The dura mater is opened in a curvilinear fashion, and the Sylvian fissure is widely dissected from distal to proximal. At the most inferior portion of the insular cortex, the insular vein and the inferior periinsular sulcus are identified (Fig. 2.1). This anatomical point corresponds to approximately 1 cm posterior to the limen insulae. A cortical incision of approximately 15 mm in length is made at this location. Gentle suction of the underlying white matter is then advanced through the cortical incision. At a depth of approximately 15 mm, the temporal horn of the lateral ventricle is reached (Fig. 2.2).

2.1.4 Resection of the Amygdala, Hippocampus, and Parahippocampal Gyrus

The inferior three-quarters of the amygdala are resected to obtain an adequate surgical corridor and to fully expose the head and body of the hippocampus (Fig. 2.3). The procedure begins with

T. Uda et al., *Atlas of Epilepsy Surgery*, https://doi.org/10.1007/978-981-92-0176-1_2

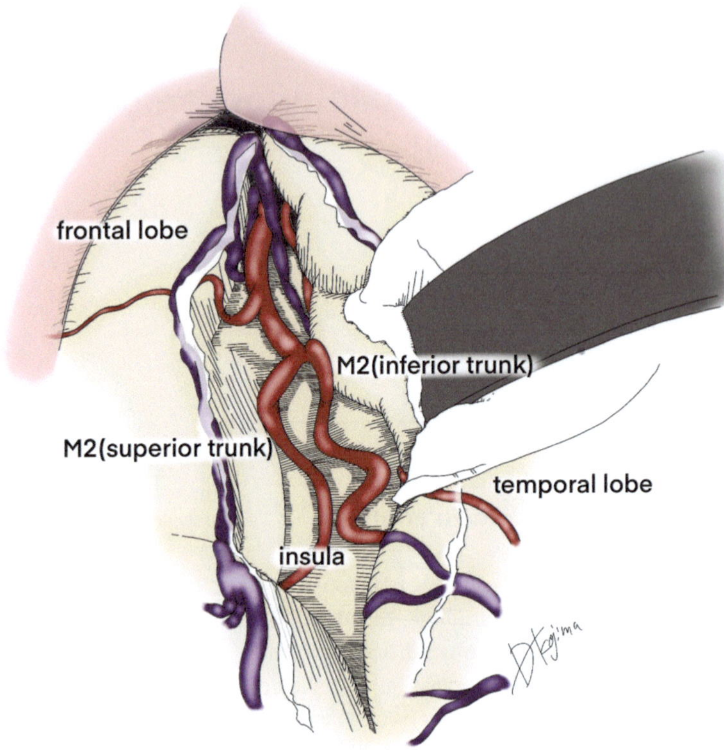

Fig. 2.1 Surgical view after Sylvian fissure dissection. After wide distal-to-proximal dissection of the Sylvian fissure, the M2 superior and inferior trunks are exposed. The insular vein and inferior periinsular sulcus are identified at the most inferior portion of the insula

the anterior resection of the uncus by suction aspiration, until the ambient cistern becomes visible through the pia mater. The inferior choroidal point is then opened.

Subsequently, the fimbria is carefully dissected away from the pia mater (Fig. 2.4). This dissection is continued posteriorly along the hippocampal sulcus. In the posteromedial region, the inferior ventricular veins draining from the hippocampus into the hippocampal sulcus are identified, coagulated, and divided (Fig. 2.5).

Laterally, dissection is initiated at the innominate sulcus (Fig. 2.6). The parahippocampal gyrus is gently elevated and separated in a subpial fashion into the operative field. Anteriorly, the hippocampal head is mobilized posteriorly, while subpial dissection is continued. During this step, hippocampal arteries arising primarily from the posterior cerebral artery (PCA) are identified, coagulated, and transected (Fig. 2.7).

Finally, disconnection is performed at the posterior portion of the hippocampal body, and the hippocampus and parahippocampal gyrus are removed en bloc (Fig. 2.8). Additional suction resection is performed to remove the posterior parahippocampal gyrus and the hippocampal tail as needed (Fig. 2.9).

2.1.5 Closure

After completion of the resection, meticulous hemostasis is confirmed. When necessary, papaverine hydrochloride is applied around the relevant vessels to prevent vasospasm. The dura is then closed, and standard wound closure is performed to complete the procedure.

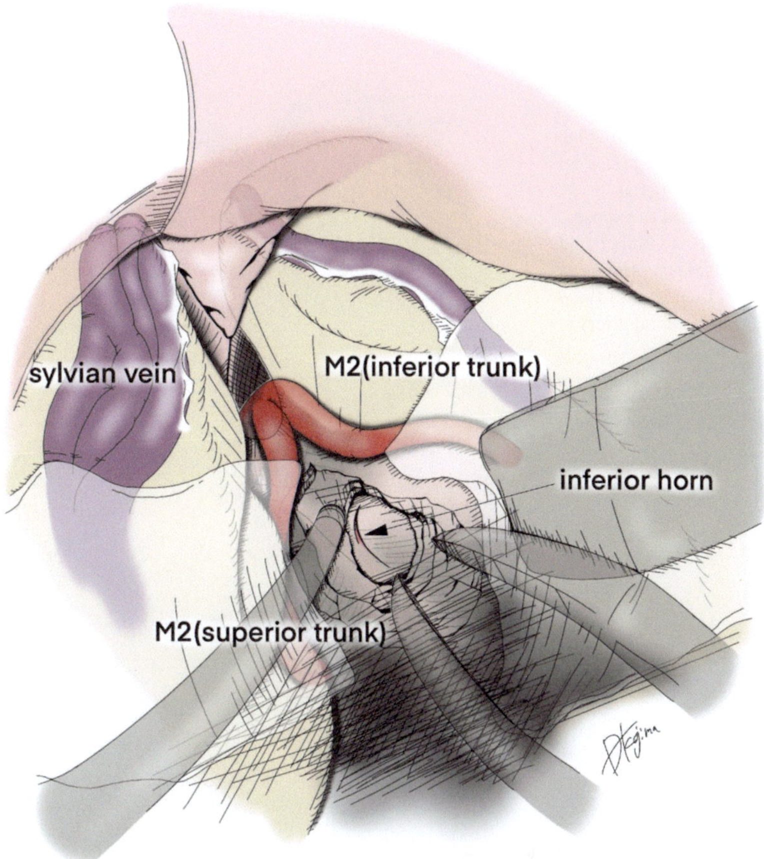

Fig. 2.2 Entry to the temporal horn. The entry point is located approximately 1 cm posterior to the limen insulae. A cortical incision of approximately 15 mm is made at this site. Gentle suction through the underlying white matter is advanced via the cortical incision, and at a depth of approximately 15 mm, the temporal horn of the lateral ventricle is reached (M2: second segment of the middle cerebral artery)

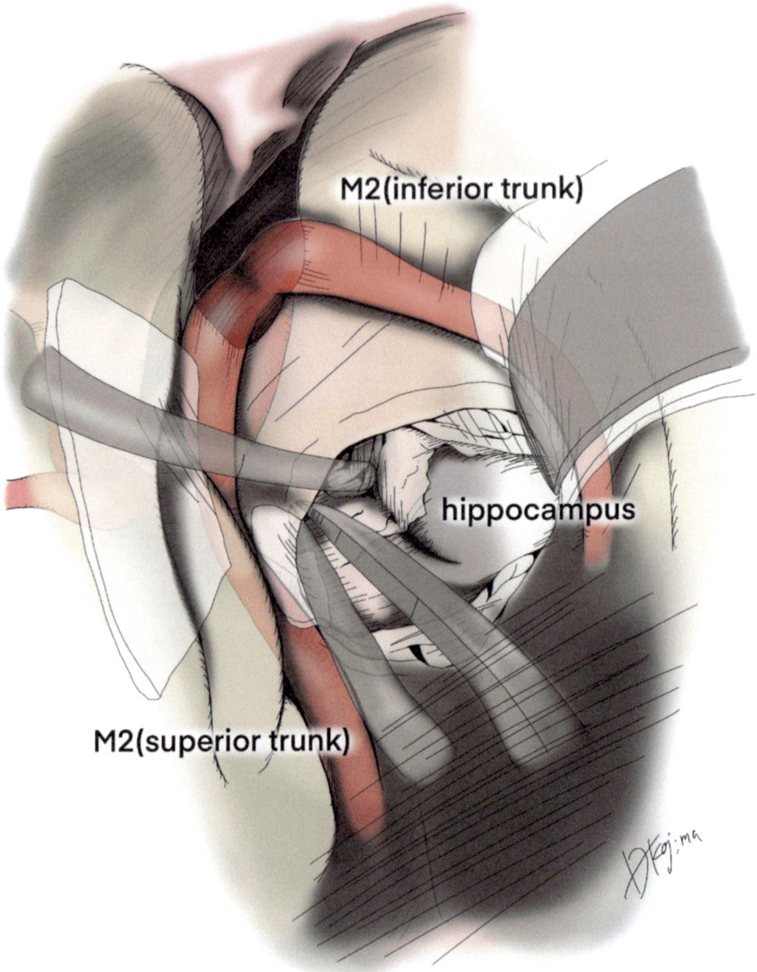

Fig. 2.3 Partial resection of the amygdala and hippocampal exposure. The inferior three-quarters of the amygdala are resected to create an adequate surgical corridor and to expose the head and body of the hippocampus (*M2* second segment of the middle cerebral artery)

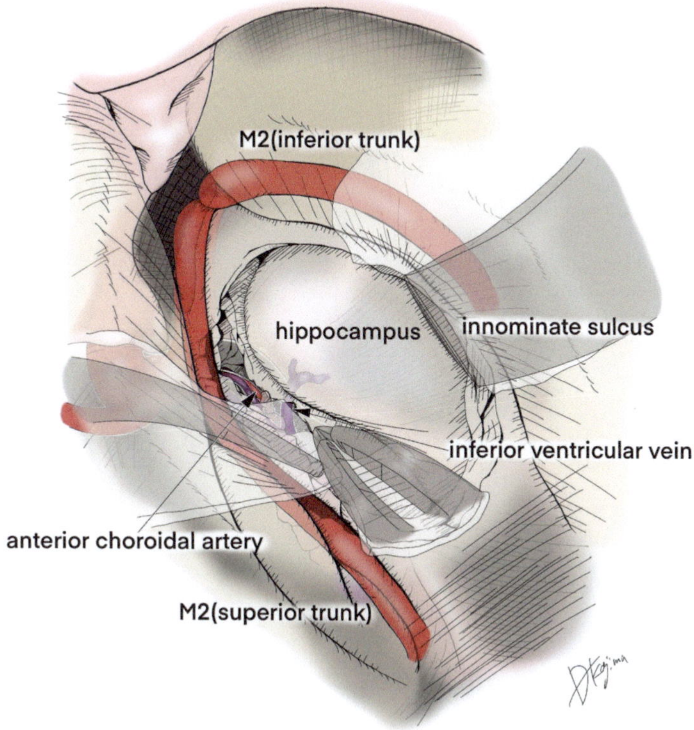

Fig. 2.4 Dissection of the fimbria. The fimbria is carefully dissected from the pia mater to mobilize the hippocampus (*M2* second segment of the middle cerebral artery)

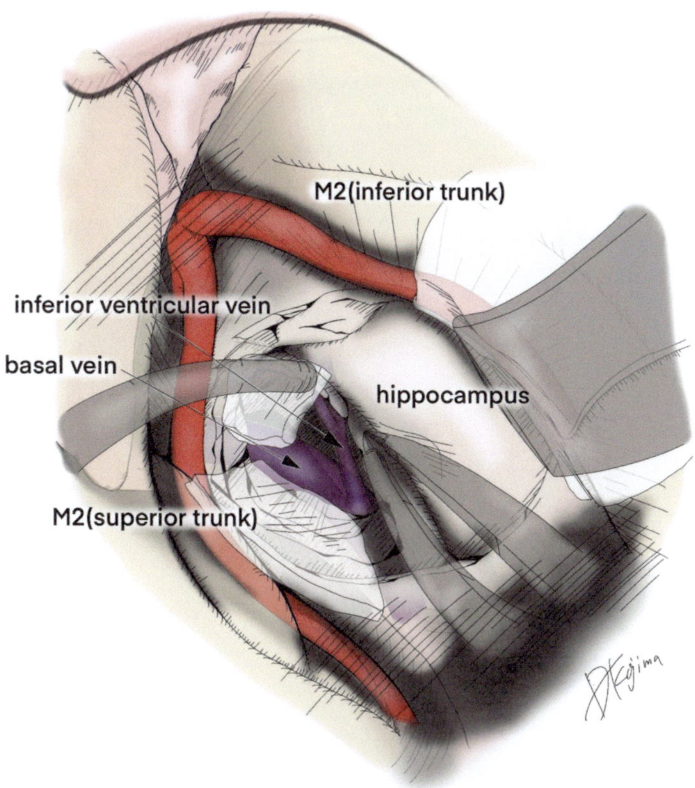

Fig. 2.5 Posteromedial dissection of the hippocampus. The dissection is continued posteriorly along the hippocampal sulcus. In the posteromedial region, inferior ventricular veins draining from the hippocampus are identified, coagulated, and divided (*M2* second segment of the middle cerebral artery)

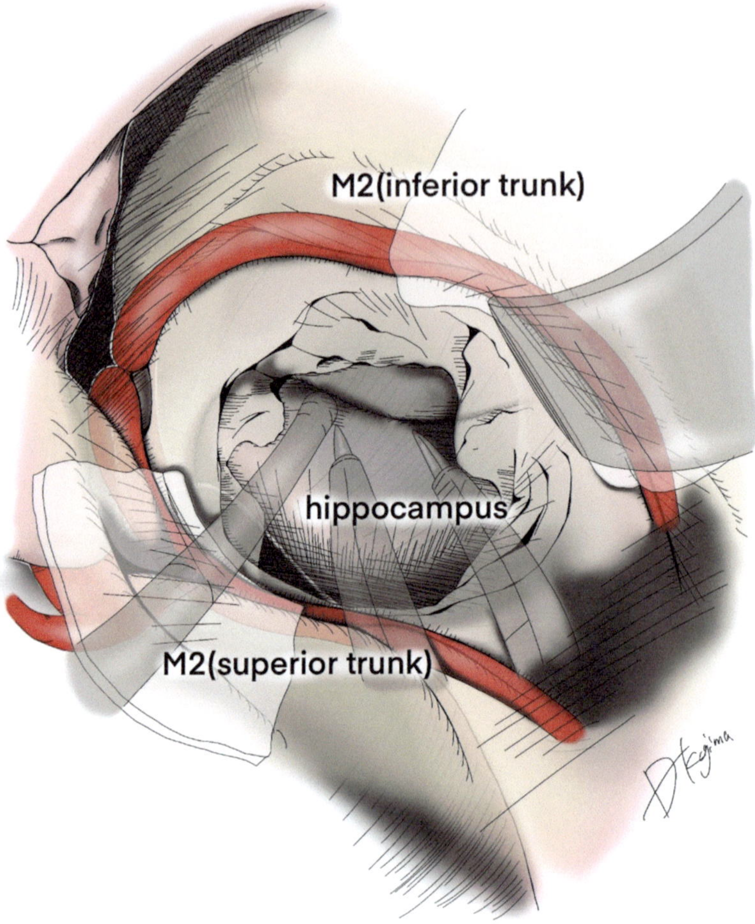

Fig. 2.6 Lateral dissection of the hippocampus and parahippocampal gyrus. Dissection is initiated laterally at the innominate sulcus. The parahippocampal gyrus is elevated subpially into the operative field (*M2* second segment of the middle cerebral artery)

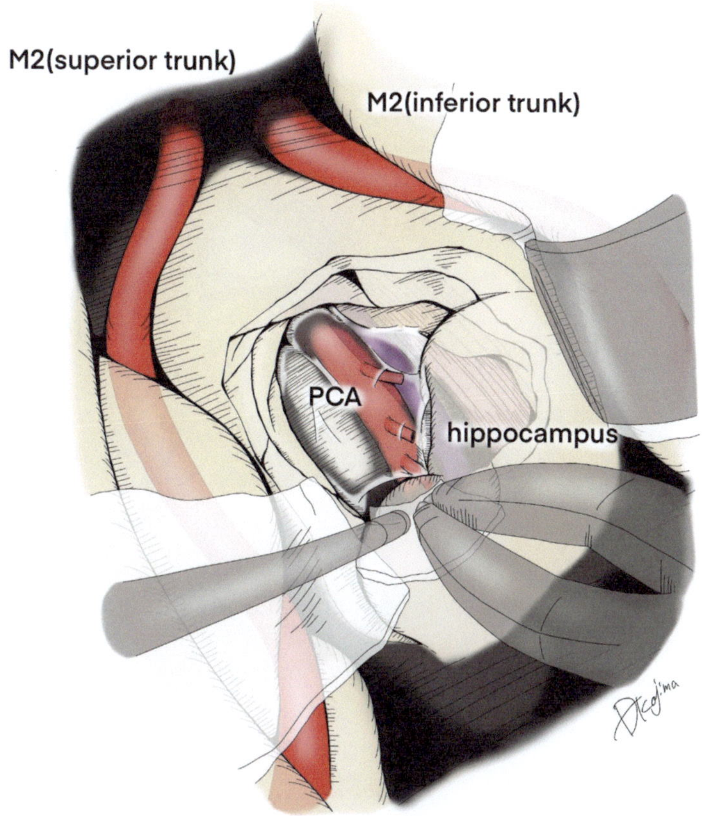

Fig. 2.7 Hippocampal mobilization with division of feeding arteries. The hippocampal head is mobilized posteriorly, while subpial dissection is continued. Hippocampal arteries arising mainly from the posterior cerebral artery (PCA) are identified, coagulated, and transected (*M2* second segment of the middle cerebral artery)

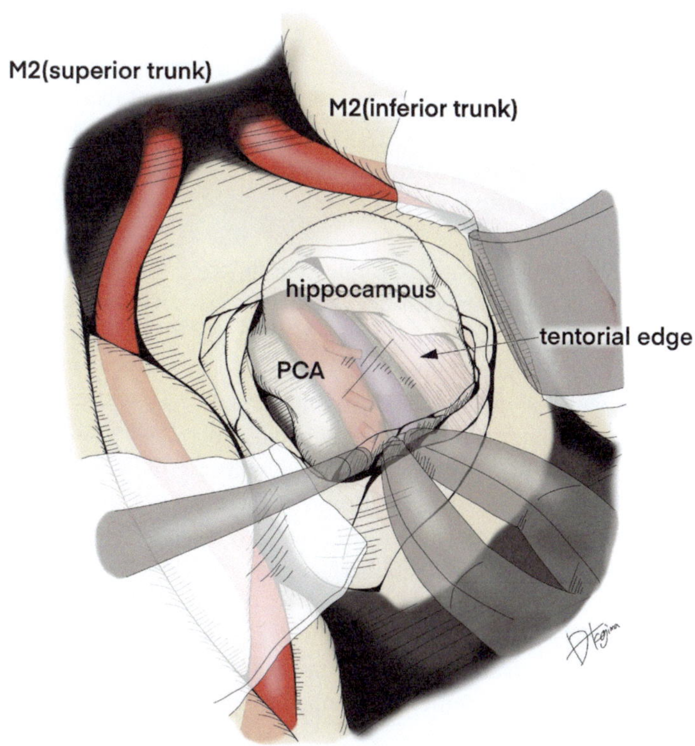

Fig. 2.8 Posterior disconnection of the hippocampal body. Disconnection is performed at the posterior portion of the hippocampal body. The hippocampus and parahippocampal gyrus are separated posteriorly in preparation for subsequent removal (*M2* second segment of the middle cerebral artery, *PCA* posterior cerebral artery)

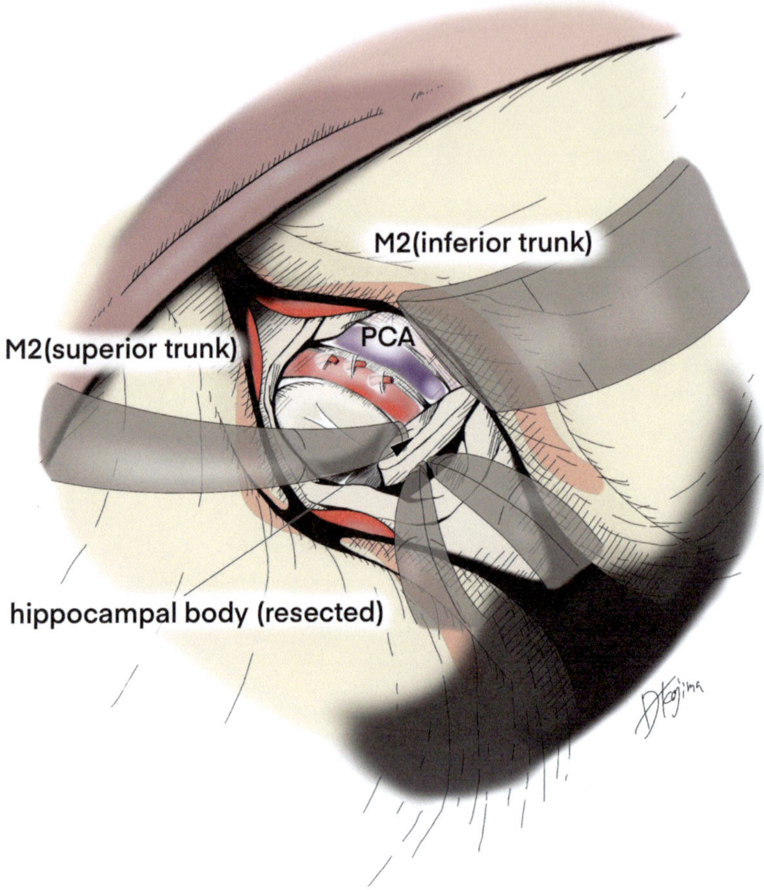

Fig. 2.9 Resection of the hippocampal tail and posterior parahippocampal gyrus. Additional suction resection is performed to remove the hippocampal tail and posterior parahippocampal gyrus (*M2* second segment of the middle cerebral artery, *PCA* posterior cerebral artery)

2.2 Note

2.2.1 Surgical Anatomy Relevant to the Transsylvian SAH

Figure 2.10 schematically illustrates the hippocampus within the temporal horn of the lateral ventricle, the fimbria, and its continuity with the fornix. In SAH, subpial dissection of the hippocampus with careful separation of the fimbria from the pia mater and continuation of this dissection into the hippocampal sulcus represents a critical surgical step. This figure is useful for understanding the three-dimensional spatial relationships of these medial temporal lobe structures.

Figure 2.11 presents a coronal view depicting the anatomical relationships among the temporal horn of the lateral ventricle, the hippocampus, and the parahippocampal gyrus. In the transsylvian SAH approach, the Sylvian fissure is dissected, the inferior periinsular sulcus is identified, and a cortical incision is made at this location to access the temporal horn. This illustration demonstrates that the surgical trajectory traverses the temporal stem.

Figure 1.12, originally presented in Chap. 1, demonstrates the vascular anatomy and cisternal structures surrounding the hippocampus. The anterior choroidal artery enters the temporal horn through the inferior choroidal point and continues into the choroid plexus. The hippocampus receives its arterial supply predominantly from branches of the posterior cerebral artery. The ambient cistern contains critical vascular and neural structures, and a thorough understanding of this anatomy is essential for the safe and effective performance of SAH.

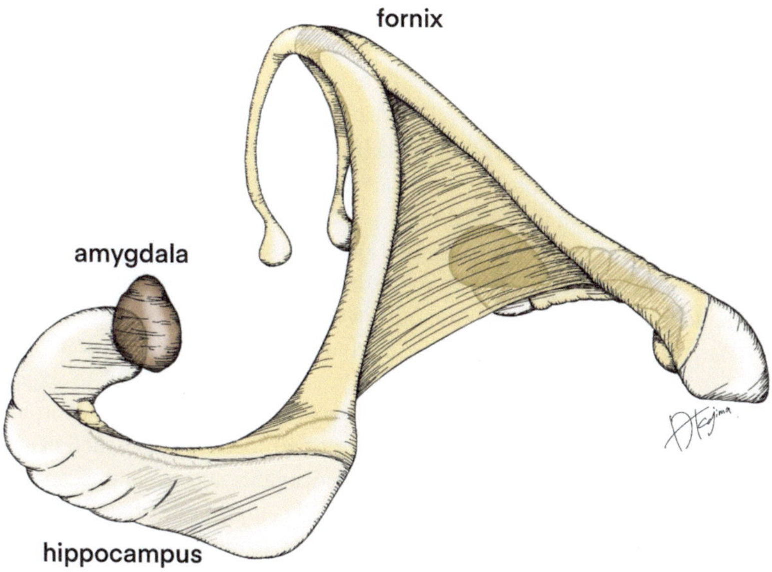

Fig. 2.10 Three-dimensional anatomy of the hippocampus, fimbria, and fornix. This schema illustrates the hippocampus within the temporal horn and its continuity through the fimbria to the fornix. Understanding these three-dimensional relationships is essential for safe subpial dissection during selective amygdalohippocampectomy

Fig. 2.11 Coronal view of the transsylvian surgical trajectory to the temporal horn. This coronal view demonstrates the transsylvian approach to the temporal horn. After the identification of the inferior periinsular sulcus, a cortical incision is made, and the surgical corridor traverses the temporal stem

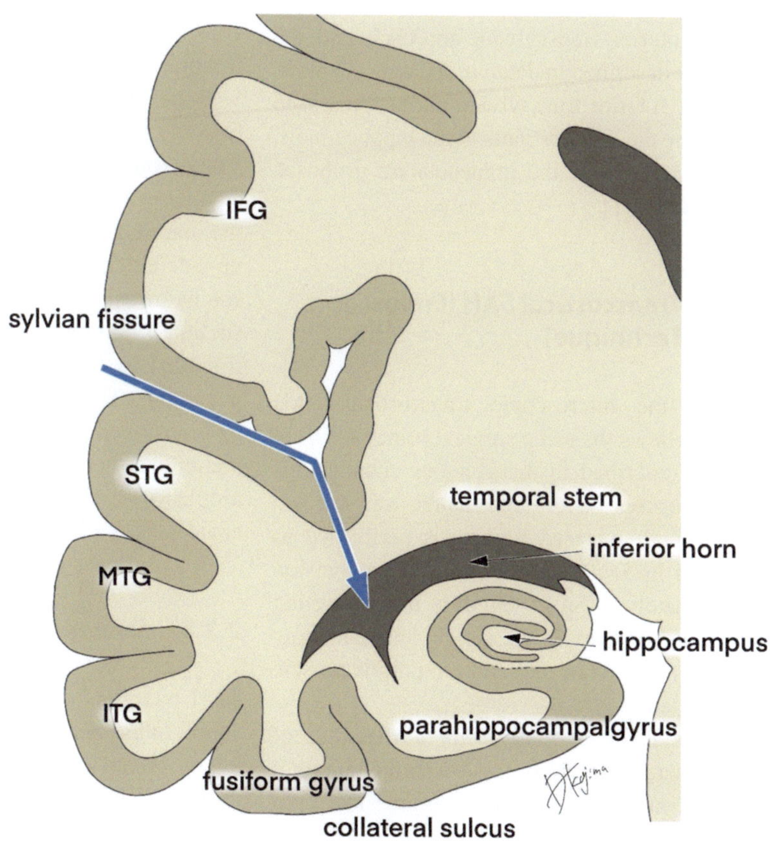

2.2.2 Five Surgical Corridors for Selective Amygdalohippocampectomy

The temporal horn of the lateral ventricle and the hippocampus are deep-seated anatomical structures that can be approached from multiple directions. In principle, selective resection of these structures can be achieved through five distinct surgical corridors, depending on the direction of access and the surrounding anatomical constraints.

These approaches include a superior route via the transsylvian approach, a lateral route via the transcortical approach, an inferior route via the subtemporal approach, an anterior route via either the anterior transsylvian approach or the trans-orbital approach, and a posterior route via the supracerebellar transtentorial approach (SCTT). Although these approaches differ substantially in their surgical trajectories and exposure, they are all directed toward the same anatomical targets, the temporal horn of the lateral ventricle and the hippocampus.

The anterior transsylvian approach and the SCTT are described in detail in separate chapters. The term anterior transsylvian approach is used to describe the anterior route to the hippocampus, in accordance with the nomenclature proposed by Türe et al. [1]

2.2.3 Transcortical SAH (Endoscopic Technique)

Because the microscopic transcortical SAH largely follows the same surgical principles as in ATL, as described in the previous chapter, this section focuses on the endoscopic transcortical SAH technique. Head fixation is performed as described in Sect. 2.1.1. A linear skin incision approximately 7 cm in length is made, starting about 2 cm anterior to the auricle and extending superiorly. The skin flap and temporalis muscle are reflected.

A circular craniotomy with a diameter of approximately 3.5 cm is created on the temporal bone lateral to the squamous suture. The dura mater is opened in a cruciate fashion. The trajectory toward the temporal horn of the lateral ventricle is confirmed from the craniotomy site, and under neuronavigation guidance, a fencepost (Nelaton catheter) is inserted from the middle temporal gyrus toward the temporal horn.

A rigid endoscope (4.0 mm diameter, 0°) is introduced, and corticectomy is advanced along the fencepost to reach the temporal horn of the lateral ventricle. Within the temporal horn, part of the amygdala located anteromedially to the hippocampus is removed to expose the hippocampal head and body. The choroid plexus is first identified medially, and suction resection of the hippocampus and parahippocampal gyrus is performed using an ultrasonic aspirator, with careful attention to preserving the pial surfaces of the basal and medial temporal lobe.

After reaching the middle cranial fossa at the level of the hippocampal head, medial suction dissection is continued toward the tentorial edge to enter the ambient cistern. At the anteromedial part of the hippocampus, the inferior choroidal point is identified, and the anterior choroidal artery and the oculomotor nerve are visualized through the pia mater.

Subsequently, suction resection is directed laterally along the dura of the middle cranial fossa from the hippocampal head. The hippocampal sulcus between the hippocampus and the parahippocampal gyrus is identified, and vessels entering the hippocampus through this sulcus are carefully coagulated. Posteromedially, suction is continued until the fimbria is reached, followed by the resection of the posterior hippocampus. The posterior extent of the resection is confirmed using neuronavigation. After meticulous hemostasis, the dura is closed, cranioplasty is performed, and the wound is closed anatomically [2].

2.2.4 Subtemporal SAH

SAH via the subtemporal approach provides access to the medial temporal lobe while preserving the lateral temporal neocortex and the temporal stem. The cortical trajectory and anatomical corridor of this approach are illustrated in Fig. 2.12. When this approach is selected, the craniotomy design and representative operative view are shown in Figs. 2.13 and 2.14.

Reported seizure outcomes vary among series; however, Engel class I outcomes are achieved in approximately 55–75% of patients, with no significant differences compared with other SAH techniques in several reports [3–5].

Although the visual field is relatively preserved with this approach, neuropsychological decline, including impairments in memory, verbal intelligence, and naming, has been reported in approximately one-third of patients undergoing surgery on the language-dominant hemisphere [2]. In contrast, Whiting et al. reported a low complication rate of 1.3% and overall favorable neuropsychological outcomes compared with other SAH approaches [4].

Despite these advantages, the subtemporal approach is associated with inherent anatomical limitations, including a deep and narrow operative corridor. Particular caution is required with respect to the tentorial edge and the venous system, especially the vein of Labbé. In addition, superiorly directed resection at the upper border of the amygdala, toward the basal ganglia, demands meticulous technique. For these reasons, although subtemporal SAH remains a valid surgical option, it is currently considered appropriate only for selected cases.

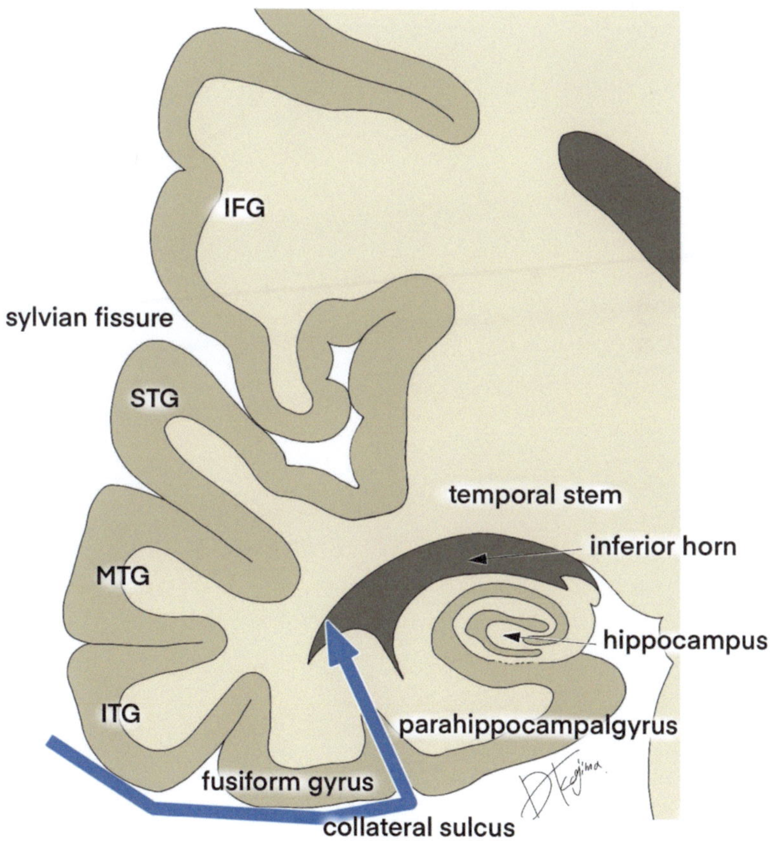

Fig. 2.12 Coronal view of the subtemporal surgical trajectory to the temporal horn. This coronal view demonstrates the subtemporal trajectory to the temporal horn. The surgical corridor advances along the basal temporal surface toward the collateral sulcus to access the inferior horn (*IFG* inferior frontal gyrus, *ITG* inferior temporal gyrus, *MTG* middle temporal gyrus, *STG* superior temporal gyrus)

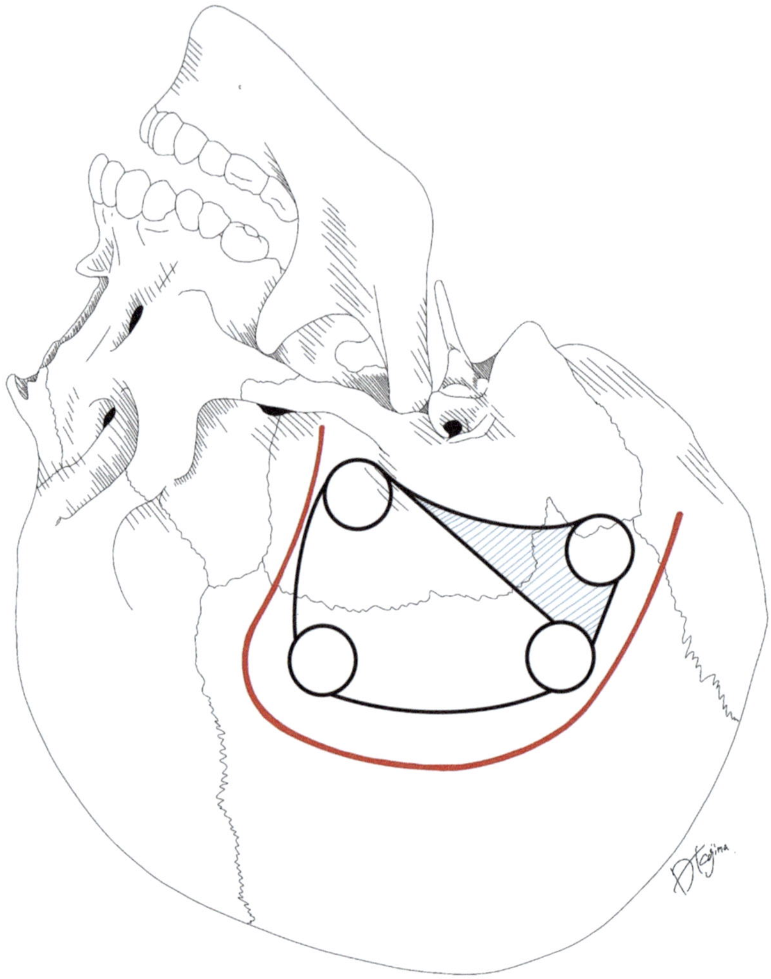

Fig. 2.13 Skin incision and craniotomy design for the subtemporal approach. The skin incision and craniotomy design for the subtemporal approach are illustrated. Burr holes are located at the asterion, around the zygomatic arch, at the supramastoid crest, and on the parietal bone. The bone flap is centered over the temporal base, providing inferior access toward the basal temporal surface and the temporal stem. The area with shaded by blue intersecting diagonal lines lines indicates the extent of bone drilling after craniotomy

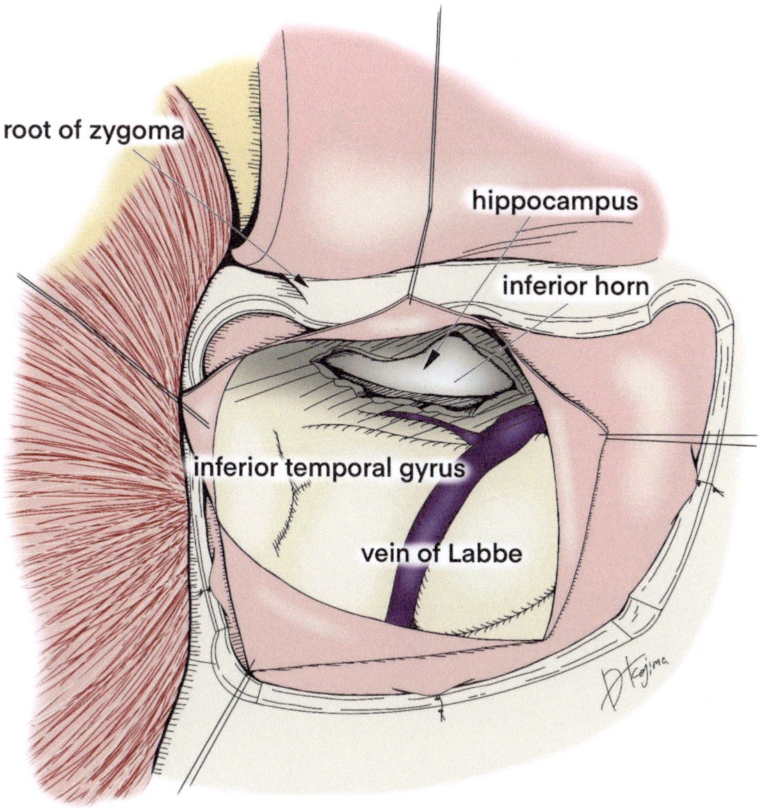

Fig. 2.14 Operative view of the subtemporal approach. Operative view following subtemporal exposure. The inferior temporal gyrus is elevated to access the inferior horn of the lateral ventricle. The hippocampus is visualized superiorly, while the vein of Labbé is preserved along the temporal surface

References

1. Dundar TT, Pehlivanoğlu MK, Eker AG, Albayrak NB, Mutluer AS, Yurtsever İ, et al. Comparison of surgical approaches to the hippocampal formation with artificial intelligence. Neurosurg Rev. 2025;48(1):251. https://doi.org/10.1007/s10143-025-03345-z.
2. Uda T, Kawashima T, Hattori M, Kojima Y, Mito Y, Goto T. Endoscopic transcortical selective amygdalohippocampectomy for mesial temporal lobe epilepsy: 2-dimensional operative video. Oper Neurosurg. 2021;21(5):E443. https://doi.org/10.1093/ons/opab265.
3. Hill SW, Gale SD, Pearson C, Smith K. Neuropsychological outcome following minimal access subtemporal selective amygdalohippocampectomy. Seizure. 2012;21(5):353–60. https://doi.org/10.1016/j.seizure.2012.03.002.
4. Whiting AC, Chen T, Swanson KI, Walker CT, Godzik J, Catapano JS, et al. Seizure and neuropsychological outcomes in a large series of selective amygdalohippocampectomies with a minimally invasive subtemporal approach. J Neurosurg. 2020;134(6):1685–93. https://doi.org/10.3171/2020.3.JNS192589.
5. Yang PF, Zhang HJ, Pei JS, Lin Q, Mei Z, Chen ZQ, et al. Neuropsychological outcomes of subtemporal selective amygdalohippocampectomy via a small craniotomy. J Neurosurg. 2016;125(1):67–74. https://doi.org/10.3171/2015.6.JNS1583.

Anterior Transsylvian Approach

3

3.1 Surgical Procedure

3.1.1 Patient Positioning and Head Fixation

The patient is placed in a supine position. With a slight vertex-down tilt, the head is rotated approximately 20° toward the contralateral side. The vertex-down position is adopted to facilitate the alignment of the operative axis with the long axis of the hippocampus by mild neck extension and to facilitate Sylvian fissure dissection.

3.1.2 Skin Incision and Craniotomy

A frontotemporal skin incision is made, followed by a right frontotemporal craniotomy. The exposure is extended until the frontozygomatic suture is clearly identified. The temporalis muscle is dissected using a two-layer technique, separating the superficial and deep fascial layers. The anterior temporal bone is sufficiently removed down to the skull base, and the sphenoid ridge is drilled down to the level where the meningo-orbital band can be identified.

3.1.3 Resection of the Hippocampus and/or Lesion

After an arcuate dural opening, the Sylvian fissure is widely dissected to secure adequate working space. The limen insulae is identified, and the course of the main trunk of the middle cerebral artery (M1) is confirmed. A cortical incision is then made at a relatively flat area of the anterior temporal lobe, located slightly deeper than this point (Fig. 3.1).

On the basal surface of the temporal lobe, the rhinal sulcus—an anterior component of the collateral sulcus that separates the parahippocampal gyrus from the fusiform gyrus—runs from the basal aspect, curves around the temporal pole, and ascends superiorly. Its superior portion can be identified within the view of a frontotemporal craniotomy. In most cases, the early temporal branch of the M1 courses along this superior segment, and the cortical incision is made medial to it (toward the Sylvian fissure).

Within the frontotemporal craniotomy field, the incision site is located slightly deeper than the limen insulae. By proceeding posteriorly with progressive white matter removal from this point, the temporal horn of the lateral ventricle can be reached approximately 1 cm posteriorly, providing a straight trajectory to the mesial temporal structures. Anatomical relationships are illustrated in Fig. 3.2.

T. Uda et al., *Atlas of Epilepsy Surgery*, https://doi.org/10.1007/978-981-92-0176-1_3

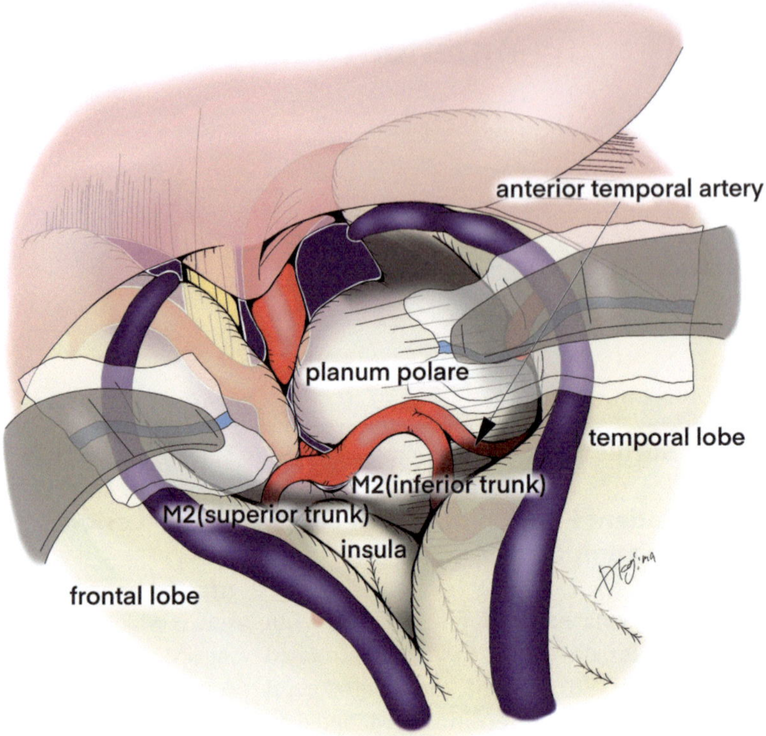

Fig. 3.1 Transsylvian exposure and anterior temporal entry. The Sylvian fissure is widely dissected to create sufficient working space. The limen insulae is identified, and the course of the main trunk of the middle cerebral artery (M1) and its bifurcation into the superior and inferior trunks (M2) are confirmed. A cortical incision is then made in a relatively flat area of the anterior temporal lobe, located slightly deeper to the limen insulae, providing access to the temporal horn

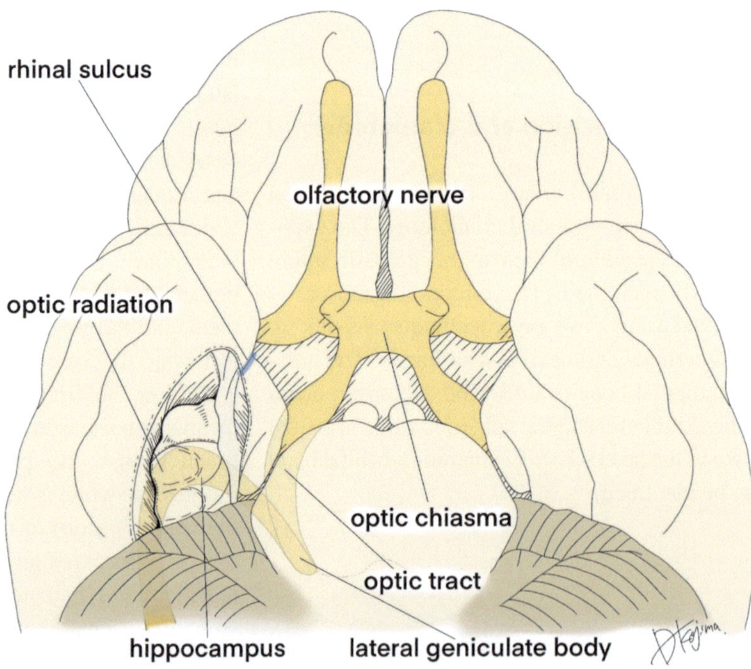

Fig. 3.2 Basal anatomical relationship between the rhinal sulcus and the temporal horn. From the basal perspective, advancing approximately 1 cm posteriorly along the medial side of the rhinal sulcus reaches the temporal horn. This surgical corridor remains remote from the lateral geniculate nucleus and Meyer's loop, minimizing the risk of visual field deficits

The extent of resection is tailored to the lesion distribution; however, this approach allows resection extending to the posterior hippocampus. During hippocampal resection, the procedure is performed while continuously identifying critical structures within the ambient cistern, including the oculomotor nerve and the anterior choroidal artery.

3.1.4 Closure

After resection, hemostasis is confirmed. If necessary, papaverine hydrochloride is applied around the relevant vessels to prevent vasospasm. The dura is closed, the wound is closed in layers, and the procedure is completed.

3.2 Note

3.2.1 Surgical Anatomy for the Anterior Transsylvian Approach

As described above, the sulcus separating the parahippocampal gyrus from the fusiform gyrus is termed the collateral sulcus, and its anterior component is referred to as the rhinal sulcus. The rhinal sulcus continues anteriorly and curves superiorly, corresponding in the frontotemporal craniotomy view to the lateral aspect of a relatively flat area located slightly deeper than the limen insulae (sometimes referred to as the planum polare). In addition, a sulcus continuous with the rhinal sulcus often contains the early temporal branch arising from the middle cerebral artery (M1), which serves as a useful landmark for identifying the starting point of the cortical incision.

Figure 3.2 presents an illustration based on the anatomical descriptions provided by Kadri et al. [1] This figure depicts the relationship between the rhinal sulcus and the temporal horn of the lateral ventricle, as viewed from the basal surface of the temporal lobe, which facilitates understanding the position of the rhinal sulcus relative to the anterior aspect of the temporal horn.

In contrast, Fig. 3.3 illustrates intraoperative findings from hippocampectomy, showing that the early temporal branch of the M1 courses along the superiorly curving portion of the rhinal sulcus. By proceeding posteriorly from the medial side of this vessel with progressive aspiration and removal of the cortical and subcortical tissue, the temporal horn of the lateral ventricle can be reached.

3.2.2 Advantages and Indications for the Anterior Transsylvian Approach

One advantage of this approach is a relatively low risk of postoperative visual field deficits. After exiting the lateral geniculate body, the Meyer loop courses anterolaterally along the roof of the temporal horn and then turns posteriorly to pass along the lateral wall of the temporal horn. As shown in Fig. 3.2, the anterior transsylvian approach reaches the temporal horn from an anteromedial direction; therefore, the surgical trajectory remains distant from the lateral geniculate nucleus and Meyer's loop, thereby reducing the risk of postoperative visual field deficits.

Accordingly, lesions localized to the uncus and anterior hippocampus, as well as cases in which avoidance of postoperative visual field deficits is a priority, are considered good candidates for this approach.

3.2.3 Another Anterior Approach to the Hippocampus or Temporal Horn: The Transorbital Approach

In addition to the anterior transsylvian approach, the transorbital approach has been reported as an anterior route to the hippocampus or the temporal horn of the lateral ventricle [2]. Endoscopic transorbital techniques have been described in recent years, offering the advantage of avoiding craniotomy while minimizing skin incision and the extent of bone removal and drilling. In contrast, because the approach traverses the orbit, the

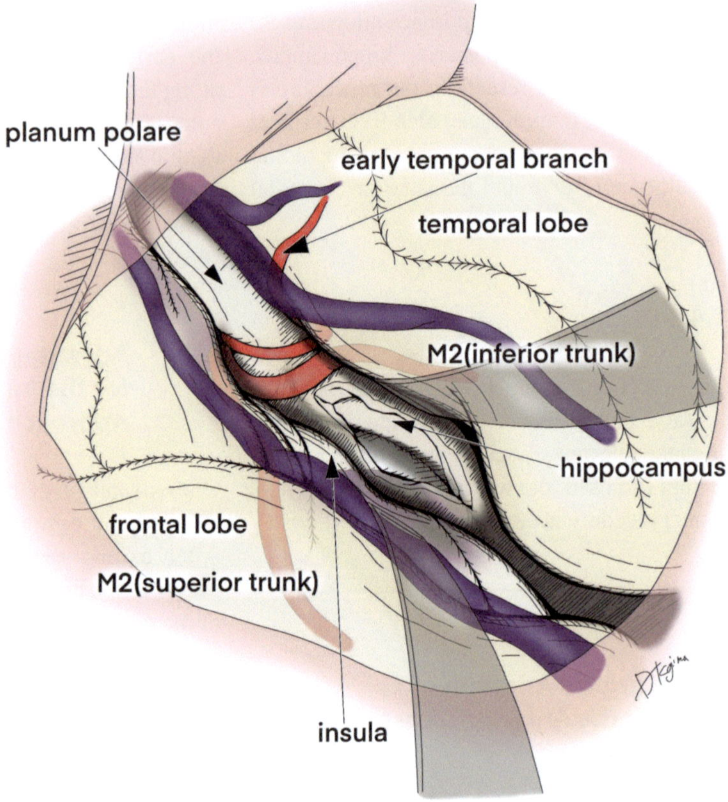

Fig. 3.3 Anatomical relationship of the rhinal sulcus, planum polare, and temporal horn. Intraoperative view demonstrating the anatomical relationship among the rhinal sulcus, the planum polare, and the temporal horn. The medial aspect of the anterior component of the rhinal sulcus corresponds to the planum polare. Advancing approximately 1 cm posteriorly along this plane leads to the temporal horn of the lateral ventricle (M2: second segment of the middle cerebral artery)

surgical corridor is narrow and the working distance tends to be long, which may limit instrument maneuverability, increase instrument interference, and restrict hemostatic control. Moreover, this approach carries orbit-specific risks, including potential injury to the optic nerve, the ophthalmic arterial system, and orbital contents. Therefore, sufficient experience in orbital and skull base surgery is required.

In contrast, although the anterior transsylvian approach requires a frontotemporal craniotomy, it provides a surgical corridor through the Sylvian fissure that affords a wide operative field and superior maneuverability. In addition, because this approach can be flexibly converted to other frontotemporal craniotomy-based procedures, such as anterior temporal lobectomy, it offers strategic versatility according to intraoperative findings.

Therefore, selecting an appropriate approach based on lesion characteristics and the surgeon's experience is important.

References

1. PAS K, de Oliveira JG, Krayenbühl N, Türe U, EPL d O, Al-Mefty O, et al. Surgical approaches to the temporal horn: an anatomic analysis of white matter tract interruption. Oper Neurosurg (Hagerstown). 2017;13(2):258–70. https://doi.org/10.1093/ons/opw011.
2. Jeon C, Hong CK, Chong K, Lee WJ, Kim GJ, Lee JI, et al. Endoscopic transorbital approach for resection of mediobasal temporal lesions using an optic radiation-sparing strategy. J Neurosurg. 2024;142(3):819–28. https://doi.org/10.3171/2024.6.JNS232810.
3. De Rosa A, Mosteiro A, Guizzardi G, Roldán P, Torales J, Matas Fassi J, et al. Endoscopic transorbital resection of the temporal lobe: anatomic qualitative and quantitative study. Front Neuroanat. 2023;17:1282226. https://doi.org/10.3389/fnana.2023.1282226.

4 Supracerebellar Transtentorial Approach (SCTT)/Paramedian Supracerebellar Transtentorial Approach (PST)

4.1 Surgical Procedure

4.1.1 Positioning and Fixation

The patient is placed in the park-bench position with the operative side up (Fig. 4.1). The head is rotated toward the contralateral side and fixed with the neck flexed (Fig. 4.2).

4.1.2 Skin Incision and Craniotomy

A hockey-stick-shaped skin incision is made in the occipital region on the operative side to expose the transverse sinus. At the midline, the incision is extended along the nuchal ligament down to the occipital bone. The suboccipital muscle group is dissected subperiosteally and reflected laterally together with the skin flap. A burr hole is placed at the asterion and another over the transverse sinus, and an additional burr hole is made inferiorly in the occipital bone to complete the craniotomy. Superiorly, the craniotomy is extended beyond the transverse sinus, while medially it is limited so as not to cross the midline. Inferiorly, the lower occipital bone is drilled away down to the level of the foramen magnum (Fig. 4.3). The dura near the foramen magnum is then opened, and the arachnoid is incised to release cerebrospinal fluid (CSF),

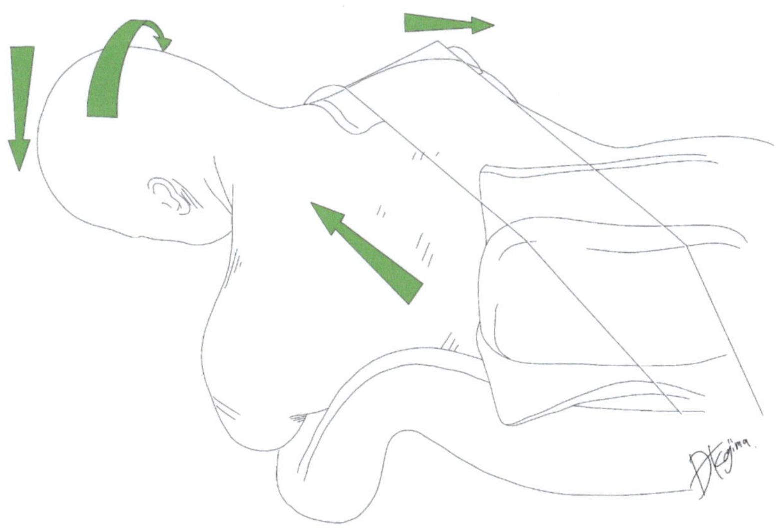

Fig. 4.1 Patient positioning in the park-bench position. The patient is placed in the park-bench position with the operative side up. The head is rotated toward the contralateral side and fixed with maximal neck flexion to optimize surgical exposure

T. Uda et al., *Atlas of Epilepsy Surgery*, https://doi.org/10.1007/978-981-92-0176-1_4

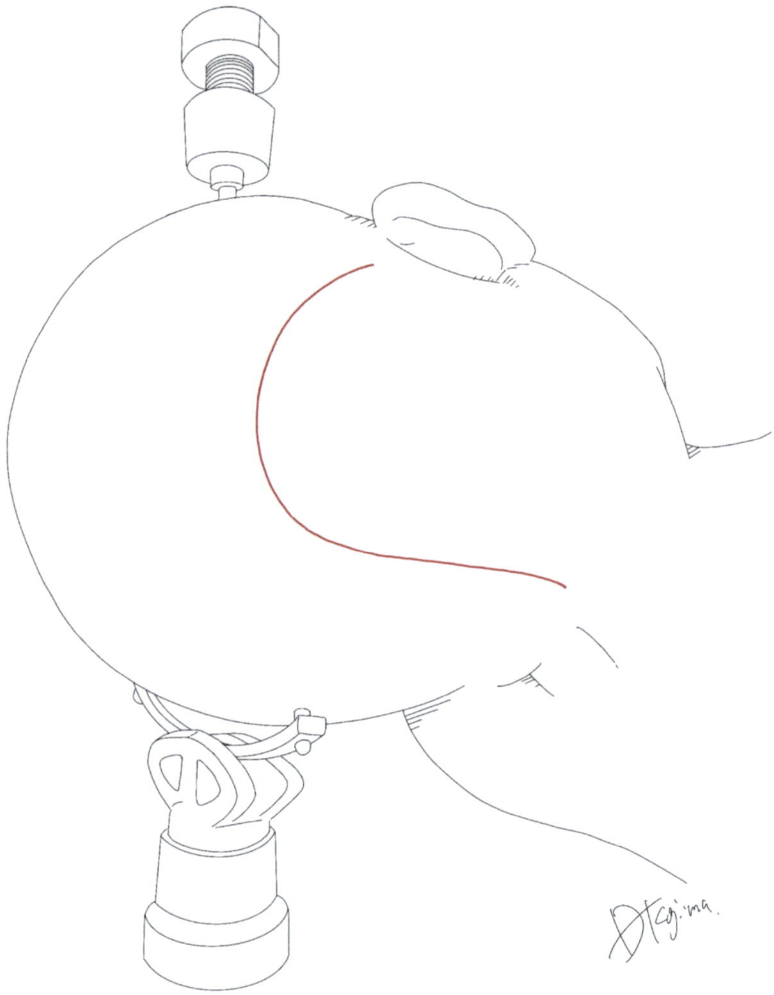

Fig. 4.2 Head rotation and fixation in the park-bench position. The head is rotated toward the contralateral side and secured with a rigid head fixation system. Adequate neck flexion and rotation are maintained to align the surgical corridor and optimize operative exposure

allowing the cerebellum to descend and create a working space beneath the tentorium. Subsequently, the dura is opened in a U-shaped manner, allowing the dural flap to be reflected superiorly toward the transverse sinus.

For intraoperative cortical electroencephalogram (EEG) evaluation, the dura can be further opened in a U-shaped manner on the supratentorial side, with the transverse sinus as the base, allowing subdural electrodes to be gently slid onto the surfaces of the parahippocampal and lingual gyri.

4.1.3 Tentorial Incision

The superior surface of the cerebellum is inspected, and the extent of the supratentorial lesion is confirmed with navigation. Among the bridging veins running from the cerebellar surface to the tentorium, those that impede cerebellar retraction or the tentorial incision are divided (Fig. 4.4). When incising the tentorium, indocyanine green (ICG) fluorescence can help identify the tentorial sinus and other vessels running within the tentorium.

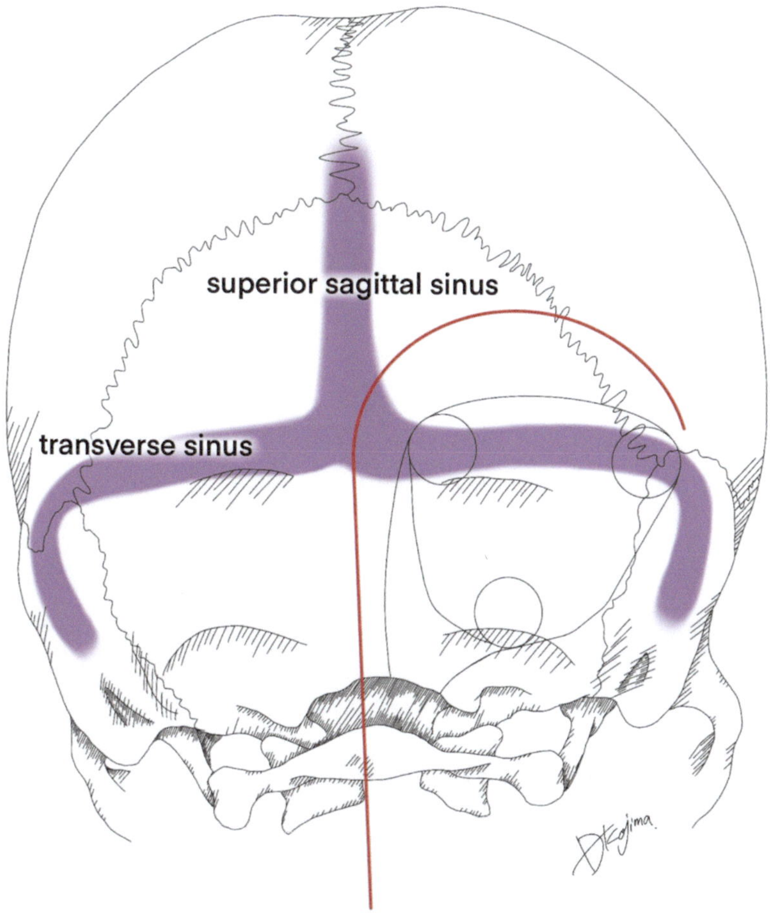

Fig. 4.3 Craniotomy design in relation to the transverse sinus. Burr holes are placed at the asterion and over the transverse sinus, with an additional inferior burr hole in the occipital bone to complete the craniotomy. The bone flap is extended superiorly beyond the transverse sinus while remaining medial to the midline. Inferiorly, the occipital bone is drilled down to the level of the foramen magnum

4.1.4 Resection of the Hippocampus and Parahippocampal Gyrus

Once the tentorium is incised, the greatest technical advantage of this approach is that it provides direct access to the lesion.

4.1.5 Closure

After confirming hemostasis, the dura is closed primarily to achieve a watertight seal. Cranial reconstruction is performed, and the wound is closed in the correct anatomical layers to avoid dead space and complete the procedure.

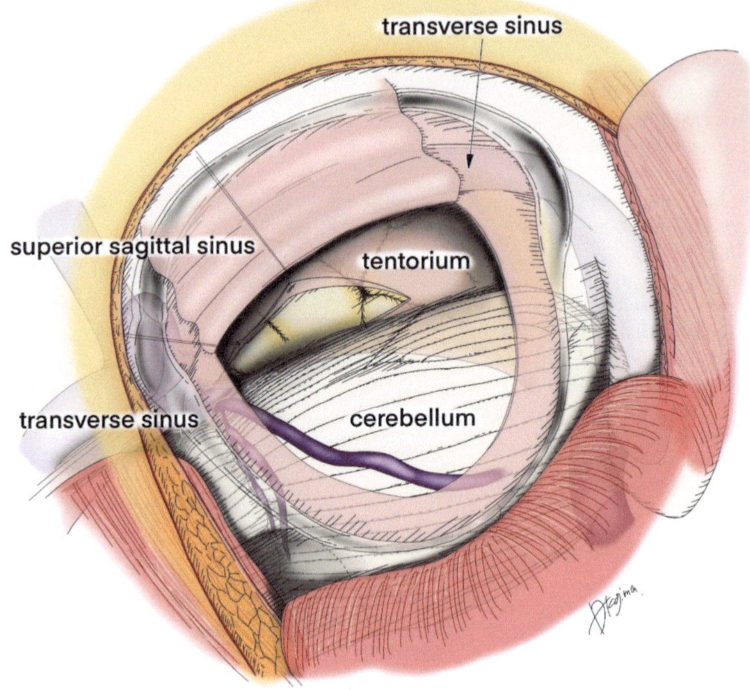

Fig. 4.4 Exposure of the basal mesial temporal lobe following tentorial incision. The tentorium is incised to expose the supratentorial compartment. This maneuver allows visualization of the basal surface of the mesial temporal lobe, including the lingual gyrus and parahippocampal gyrus

4.2 Notes

4.2.1 Surgical Anatomy Required for the SCTT Approach

As described above, once the tentorium is incised, the surfaces of the medial temporal structures become directly visible, which is a key element of this approach. An illustration adapted from the report by de Oliveira et al. is shown in Fig. 4.5 [1].

In this approach, one often encounters bridging veins between the cerebellar surface and the tentorium that may interfere with surgical manipulation. Yonekawa et al. reported that approximately one-third of cases required coagulation and division of a large bridging vein to complete this approach. No clinical complications were observed after sacrificing these veins, although complications have been reported by other authors following similar maneuvers [2]. Türe et al. emphasized the importance of preserving vermian bridging veins, as injury to these veins may result in cerebellar edema, venous infarction, or hemorrhage. They also noted that a few small bridging veins are present on the paramedian superior surface of the cerebellum beneath the tentorium. In their early four cases (26.7% of the series), these small veins were sacrificed to obtain an adequate operative field, and no clinical or neuroradiological complications were observed [3]. Vermian veins may constitute a major drainage pathway into the deep venous system, including the vein of Galen. Therefore, even when a bridging vein is to be divided, careful assessment is mandatory.

4.2.2 Indications, Patient Positioning, and Comparison with Other Approaches in SCTT

The most notable advantage of this approach is its low risk of postoperative visual field deficits [4]. Regarding the selection of surgical

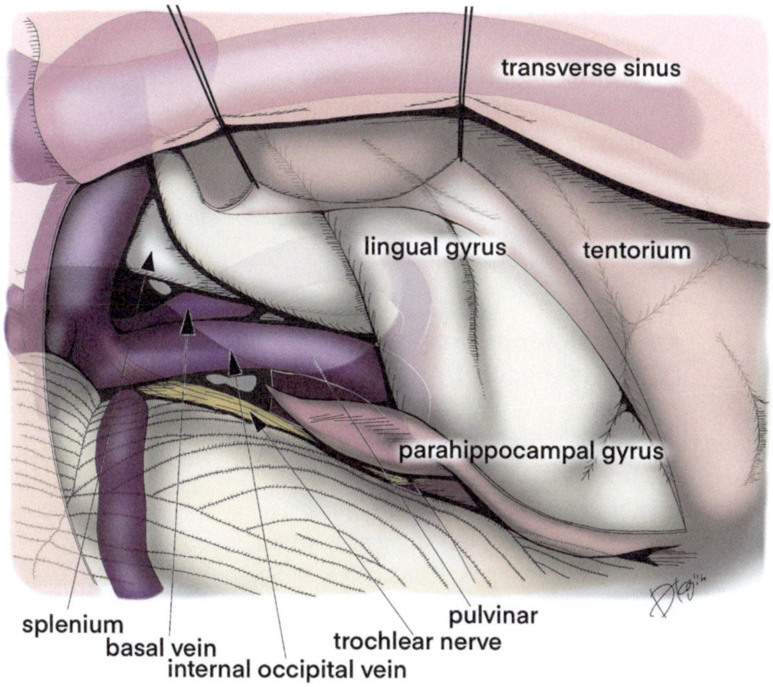

Fig. 4.5 Surgical view in the SCTT approach. Following tentorial incision in the supracerebellar transtentorial (SCTT) approach, the medial temporal region is accessed from below. This illustration depicts the lingual gyrus, parahippocampal gyrus, and adjacent deep structures to demonstrate the anatomical corridor of the SCTT approach

approaches based on lesion location, Türe et al. recommended the transsylvian approach for lesions confined to the anterior portion of the medial temporal region or extending from the anterior to the mid-portion. In contrast, for lesions confined to the mid- or posterior portion of the medial temporal region, or extending from the mid- to the posterior portion, they recommended the SCTT approach [3]. They noted that the SCTT approach can also be applied to lesions involving the entire medial temporal region, although this largely depends on the surgeon's experience. A report classifies the anterior, middle, and posterior portions of the medial temporal region as follows: the anterior portion is separated from the middle portion by the posterior margin of the uncus, and the middle portion is separated from the posterior portion at the level of the colliculi, corresponding to the anterior margin of the splenium of the corpus callosum [5].

Regarding patient positioning for the SCTT approach, there are reports describing the use of the semi-sitting position, the Concorde position, and the park-bench position. In the semi-sitting position, gravity allows the cerebellum to descend inferiorly, thereby widening the surgical corridor; however, the approach toward the temporal lobe through the tentorium becomes angled from inferior to superior, and there is also a risk of air embolism. In the Concorde position, the operative field lies close to the level of the heart, which can reduce venous return and lead to higher venous pressures. The distance from the surgeon to the operative field also tends to be greater than in other positions. In the park-bench position we use, the superior surface of the cerebellum and the tentorium are easily visualized, and the distance from the surgeon to the operative field is relatively short. Although the cerebellum does not descend as easily as in the semi-sitting position, sufficient working space can be obtained by opening the foramen magnum and releasing CSF.

Compared with the subtemporal approach, the subtemporal route requires greater retraction of the temporal lobe, which may increase the risk of contusion. In addition, excessive retraction may injure major draining veins such as the vein of Labbé, leading to venous infarction; this can result in severe neurological deficits, particularly in the dominant hemisphere. The subtemporal approach provides an operative field that is, on average, 1.8 cm deeper than that obtained with the SCTT approach [6]. In contrast, the SCTT approach reaches the medial temporal structures through a posterior supracerebellar transtentorial trajectory, allowing access without traversing brain parenchyma and along a more favorable, less steep operative angle.

4.2.3 Selective Amygdalohippocampectomy (SAH) via the SCTT Approach

Reports of selective amygdalohippocampectomy (SAH) performed via the SCTT approach are limited. Serra et al. reported a series of 17 patients who underwent SAH using this approach, with 76.4% achieving Engel class I outcomes at a mean follow-up of 9 months [4]. No complications related to the semi-sitting position and no visual field deficits were observed.

References

1. de Oliveira JG, Párraga RG, Chaddad-Neto F, Ribas GC, de Oliveira EP. Supracerebellar transtentorial approach: resection of the tentorium instead of an opening to provide broad exposure of the mediobasal temporal lobe—anatomical aspects and surgical applications: clinical article. J Neurosurg. 2012;116(4):764–72. https://doi.org/10.3171/2011.12.JNS111256.
2. Yonekawa Y, Imhof HG, Taub E, Curcic M, Kaku Y, Roth P, et al. Supracerebellar transtentorial approach to posterior temporomedial structures. J Neurosurg. 2001;94(2):339–45. https://doi.org/10.3171/jns.2001.94.2.0339.
3. Türe U, Harput MV, Kaya AH, Baimedi P, Firat Z, Türe H, et al. The paramedian supracerebellar-transtentorial approach to the entire length of the mediobasal temporal region: an anatomical and clinical study. J Neurosurg. 2012;116(4):773–91. https://doi.org/10.3171/2011.12.JNS11791.
4. Serra C, Akeret K, Staartjes VE, Ramantani G, Grunwald T, Jokeit H, et al. Safety of the paramedian supracerebellar-transtentorial approach for selective amygdalohippocampectomy. Neurosurg Focus. 2020;48(4):E4. https://doi.org/10.3171/2020.1.FOCUS19909.
5. Weil AG, Middleton AL, Niazi TN, Ragheb J, Bhatia S. The supracerebellar-transtentorial approach to posteromedial temporal lesions in children with refractory epilepsy. J Neurosurg Pediatr. 2015;15(1):45–54. https://doi.org/10.3171/2014.10.PEDS14162.
6. Chau AMT, Gagliardi F, Smith A, Pelzer NR, Stewart F, Mortini P, et al. The paramedian supracerebellar transtentorial approach to the posterior fusiform gyrus. Acta Neurochir. 2016;158(11):2149–54. https://doi.org/10.1007/s00701-016-2960-8.

Hippocampal Transection

5

5.1 Surgical Procedure

5.1.1 Positioning, Head Fixation, and Craniotomy

The procedures from skin incision to craniotomy are the same as those used in selective amygdalohippocampectomy via the transsylvian approach. Briefly, the patient is placed in the supine position, with the head rotated approximately 20° away from the surgeon. A standard frontotemporal craniotomy is performed; however, slightly wider exposure of the temporal lobe is generally preferred.

5.1.2 Sylvian Fissure Dissection and Identification of the Temporal Horn

The procedures at this stage are largely comparable to those described for transsylvian selective amygdalohippocampectomy in Chap. 2. In brief, after the wide opening of the Sylvian fissure, the inferior periinsular sulcus is identified. A cortical incision of approximately 15 mm is made along this sulcus to enter the temporal horn of the lateral ventricle, with preservation of the uncinate fasciculus running anterior to the limen insulae.

After entering the temporal horn of the lateral ventricle, the hippocampus is identified along its floor, with a portion of the amygdala visualized facing it. To secure an adequate operative field for subsequent procedures, partial resection of the inferior portion of the amygdala and the superior wall of the temporal horn is performed. This maneuver provides sufficient exposure from the hippocampal head to the body within the temporal horn, allowing a clear identification of the choroid plexus and the fimbria.

Subsequently, intraoperative electrocorticography is performed from the hippocampal surface to evaluate epileptiform discharges originating from the hippocampus.

5.1.3 Hippocampal Transection

Hippocampal transection (HT) is performed in a direction perpendicular to the long axis of the hippocampus. First, the alveus is incised using microscissors. A 5-mm round knife is then introduced through this incision to transect the pyramidal cell layer located within 2 mm from the surface of the hippocampus. At the deepest extent of the incision, arterial and venous structures running within the hippocampal sulcus can be observed through the arachnoid membrane.

In this procedure, the preservation of the fimbria is essential. The fimbria is a fiber bundle running parallel to the long axis of the hippocampus and plays an important role in memory-related pathways. This maneuver is repeated four to six times at approximately 5-mm intervals, progressing from the hippocampal head toward the posterior aspect of the body (Fig. 5.1).

T. Uda et al., *Atlas of Epilepsy Surgery*, https://doi.org/10.1007/978-981-92-0176-1_5

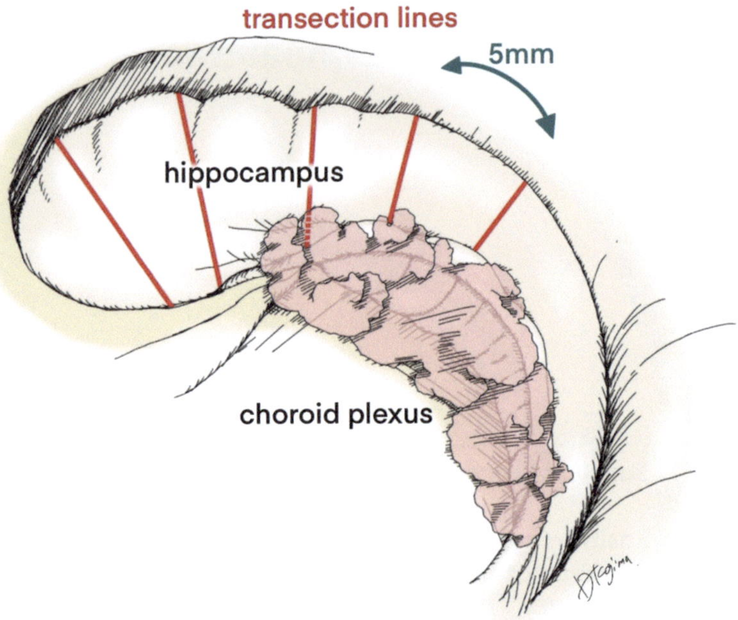

Fig. 5.1 Hippocampal transection. Hippocampus is transected in a direction perpendicular to the long axis of the hippocampus. This maneuver is repeated four to six times at approximately 5-mm intervals, progressing from the hippocampal head toward the posterior aspect of the body

5.1.4 Parahippocampal Gyrus Transection

A 4-mm multiple subpial transection (MST) hook is introduced along the previously created incision, using the innominate sulcus as a landmark, and advanced to transect the parahippocampal gyrus and the entorhinal cortex.

The entorhinal cortex is considered to play a key role in the propagation of epileptiform discharges and is therefore an important target for interrupting epileptic networks. Because this maneuver inevitably involves blind manipulation, rough handling carries a risk of injury to critical vascular structures within the ambient cistern and to the midbrain. For this reason, a blunt-tipped MST hook is used, and the maneuver is performed in a controlled and gentle manner to avoid excessive penetration (Fig. 5.2).

5.1.5 Posttransection Assessment

After completion of multiple transections of the hippocampus and parahippocampal gyrus, intraoperative electrocorticography is repeated from the hippocampal surface. The disappearance or marked attenuation of preoperatively identified abnormal discharges is evaluated to assess whether adequate disconnection has been achieved.

5.1.6 Closure

After confirming adequate hemostasis, the dura mater is closed, followed by standard cranial and wound closure to complete the procedure.

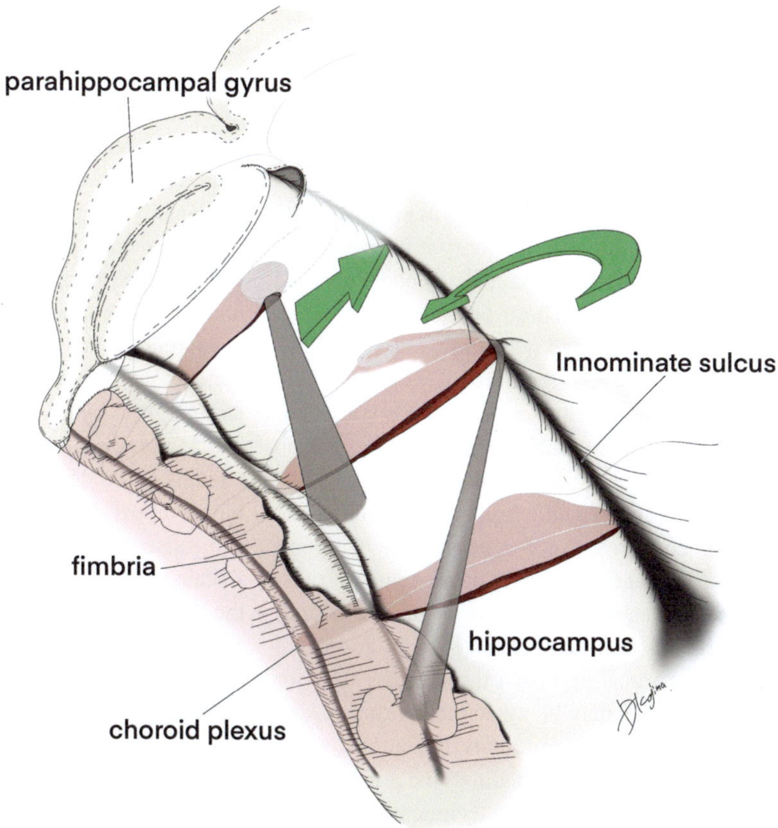

Fig. 5.2 Each procedure of hippocampal transection. The alveus is incised using microscissors. At the deepest extent of the incision, arterial and venous structures running within the hippocampal sulcus can be observed, preserving the fimbria. Then a blunt hook is introduced along the previously created incision, using the innominate sulcus as a landmark, and advanced to transect the parahippocampal gyrus and the entorhinal cortex

5.2 Note

5.2.1 Seizure Outcome

Hippocampal transection (HT), unlike anterior temporal lobectomy (ATL) or selective amygdalohippocampectomy (SAH), is a surgical technique aimed at interrupting seizure propagation rather than removing epileptogenic tissue. Regarding seizure outcomes, Patil et al. reported that 14 of 15 patients (94.7%) who underwent HT and were followed for more than 2 years achieved an Engel class I outcome [1]. In addition, a systematic review of 59 patients treated with HT, either as a stand-alone procedure or in combination with other surgical techniques, reported a seizure freedom rate of 75% at 12 months or longer after surgery [2]. HT has been reported in a relatively small number of cases compared with SAH or ATL, and large cohort studies or prospective trials are scarce. Consequently, a direct comparison of seizure outcomes between HT and SAH or ATL remains difficult at present. Although HT has demonstrated a certain degree of seizure control as a promising minimally invasive procedure, robust evidence regarding long-term seizure outcomes remains limited, and further case accumulation as well as comparative studies are warranted.

5.2.2 Complications

HT is a surgical technique designed to minimize language and memory impairments, particularly in procedures involving the dominant hemisphere. By creating transections perpendicular to the long axis of the hippocampus, HT selectively interrupts longitudinal intrahippocampal fibers while preserving fiber connections running in the orthogonal direction. As a result, seizure propagation can be suppressed while maintaining the functional continuity of the hippocampus.

Regarding neuropsychological outcomes, Patil et al. reported improvement in verbal memory in seven of nine evaluable patients, with no change observed in the remaining two [1]. In a systematic review by Abramov et al., postoperative verbal memory improved in 16 of 38 patients (42%) and remained unchanged in eight patients (21%), while nonverbal memory improved in 12 patients (32%) and remained unchanged in 12 patients (32%) [2]. These findings suggest that HT may be associated with a low risk of postoperative memory deterioration.

To date, few serious postoperative complications directly attributable to HT itself have been reported. However, complications related to the surgical approach to the hippocampus may pose clinical concerns. Approaches to the temporal horn of the lateral ventricle have included transcortical routes via the superior or middle temporal gyrus, as well as the transsylvian approach. The transcortical approach through the superior temporal gyrus may be associated with a risk of postoperative language deficits. In contrast, while the transsylvian approach allows the preservation of the temporal cortex, it is technically demanding, and sectioning of the temporal stem may risk injury to Meyer's loop, leading to postoperative hemianopia, as reported in previous studies [3].

We evaluated 37 patients with mesial temporal lobe epilepsy without hippocampal sclerosis or evident hippocampal atrophy who underwent transsylvian HT. Postoperatively, verbal memory showed significant improvement in patients who underwent right-sided surgery, whereas no significant pre-to-postoperative change was observed in those who underwent left-sided surgery [4]. These findings suggest that HT may be a useful function-preserving surgical option when hemispheric dominance is taken into account while also indicating that careful interpretation is required when considering its effects in the dominant hemisphere.

5.2.3 SEEG-Guided RF-MHT: An Extension of HT

A minimally invasive technique based on the principles of hippocampal transection, stereo-electroencephalography-guided radiofrequency multiple hippocampal transection (SEEG-guided RF-MHT), has been described in recent studies. A major advantage of this technique is that it does not require craniotomy. In SEEG-guided RF-MHT, radiofrequency ablations are delivered through implanted SEEG electrodes, and three linear lesions oriented perpendicular to the long axis of the amygdalo-hippocampal complex are created under awake conditions at the patient's bedside.

García Losarcos et al. reported a series of five patients who underwent SEEG-guided RF-MHT targeting the language-dominant hemisphere, in which seizure freedom was achieved in four patients (80%) [5]. At 6 months postoperatively, hippocampal volume was well preserved, and no subjective memory complaints were reported. Because this technique allows the preservation of the temporal stem, SEEG-guided RF-MHT may represent a treatment option with favorable cognitive outcomes, and further development and validation of this approach are anticipated.

References

1. Patil AA, Andrews R. Long-term follow-up after multiple hippocampal transection (MHT). Seizure. 2013;22(9):731–4. https://doi.org/10.1016/j.seizure.2013.05.014.
2. Abramov I, Jubran JH, Houlihan LM, Park MT, Howshar JT, Farhadi DS, et al. Multiple hippocampal transection for mesial temporal lobe epilepsy: a systematic review. Seizure. 2022;101:162–76. https://doi.org/10.1016/j.seizure.2022.08.007.
3. Pitskhelauri D, Kudieva E, Kamenetskaya M, Kozlova A, Vlasov P, Dombaanai B, et al. Multiple hippocampal transections for mesial temporal lobe epilepsy. Surg Neurol Int. 2021;12:372. https://doi.org/10.25259/SNI_350_2021.
4. Uda T, Morino M, Ito H, Minami N, Hosono A, Nagai T, et al. Transsylvian hippocampal transection for mesial temporal lobe epilepsy: surgical indications, procedure, and postoperative seizure and memory outcomes. J Neurosurg. 2013;119(4):1098–104. https://doi.org/10.3171/2013.6.JNS13244.
5. Garcia Losarcos N, Miller J, Fastenau P, Amina S, Bailey C, Deveraux M, et al. Stereotactic-EEG-guided radiofrequency multiple hippocampal transection (SEEG-guided-RF-MHT) for the treatment of mesial temporal lobe epilepsy: a minimally invasive method for diagnosis and treatment. Epileptic Disord. 2021;23(6):905–15. https://doi.org/10.1684/epd.2021.1314.

Corpus Callosotomy 6

6.1 Surgical Procedure of Total/ Anterior Corpus Callosotomy

6.1.1 Patient Position, Skin Incision, and Craniotomy

The corpus callosum is divided into five parts: rostrum, genu, body, isthmus, and splenium. Corpus callosotomy is generally classified into anterior two-thirds callosotomy, posterior one-third callosotomy, and total callosotomy. The isthmus serves as a border of anterior two-thirds and posterior one-third callosotomy. Total and anterior corpus callosotomy are performed via an anterior approach, whereas posterior corpus callosotomy is performed via a posterior approach (Fig. 6.1).

In this session, we separately describe the surgical procedures.

Most surgeons place the patient in the supine position without head rotation, while others prefer to set the patient in a lateral position with the entry side down to facilitate dissection of the interhemispheric fissure with the aid of gravity (Fig. 6.2). A coronal skin incision is used, and the craniotomy is approximately 6 cm in diameter (Fig. 6.3). The coronal suture serves as the posterior margin of the craniotomy, and a frontal craniotomy is created extending approximately 1 cm beyond the midline.

In most cases, craniotomy around the central sulcus should be avoided because the largest bridging vein to the superior sagittal sinus (SSS), the so-called vein of Trolard, often courses near this region. The side of entry into the interhemispheric fissure is also determined by the bridging veins, favoring the side with fewer obstructing veins. Preoperative venous imaging is useful for identifying bridging veins to the SSS. Noncontrast magnetic resonance venography (MRV) is often sufficient, while contrast-enhanced computed tomography (CT) or MRI venography may also be helpful when additional information is required.

6.1.2 Interhemispheric Fissure Dissection

The dura is opened toward the midline (Fig. 6.4). Then the frontal lobe is gently retracted laterally. Large bridging veins from the frontal lobe to the superior sagittal sinus (SSS) should be preserved as much as possible. Along the medial surface of the superior frontal gyrus, the falx cerebri separates the hemispheres. This structure ends at the level of the cingulate gyrus, and the hemispheres are tightly apposed below the inferior edge of the falx cerebri. To approach the corpus callosum without injuring the cerebral parenchyma, meticulous dissection of the interhemispheric fissure is required. Sharp dissection should be performed carefully to avoid injury to the pia mater. Drainage of cerebrospinal fluid (CSF) from the arachnoid

T. Uda et al., *Atlas of Epilepsy Surgery*, https://doi.org/10.1007/978-981-92-0176-1_6

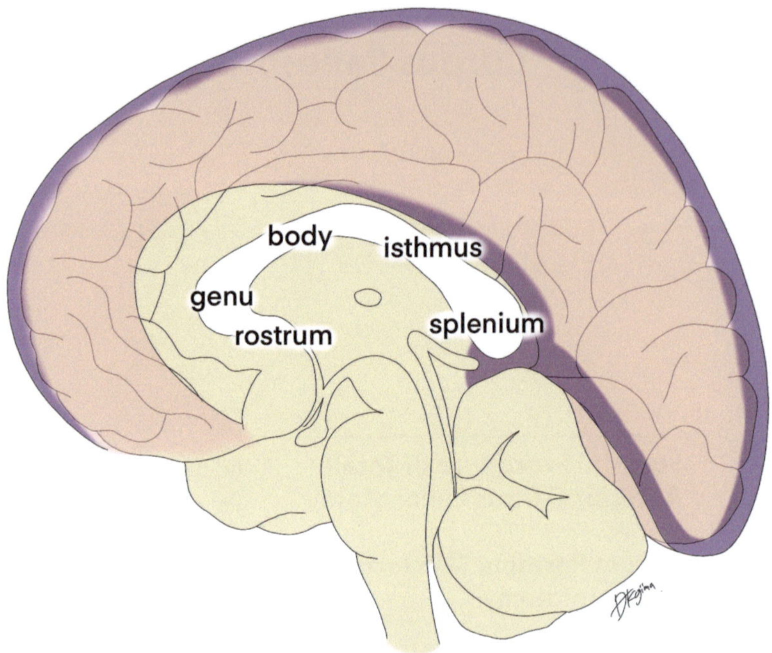

Fig. 6.1 Illustration of the corpus callosum, vein, and falx. Corpus callosum is divided into five parts: rostrum, genu, body, isthmus and splenium

space facilitates brain relaxation. For effective retraction, the position of the spatula should be adjusted repeatedly as needed.

Within the interhemispheric fissure, the callosomarginal arteries and then the pericallosal arteries are identified. The body of the corpus callosum is identified between the bilateral pericallosal arteries (Fig. 6.5).

6.1.3 Severing the Corpus Callosum

The three-dimensional anatomical relationships among the corpus callosum, fornix, and lateral ventricle are demonstrated (Fig. 6.6). To prevent hydrocephalus and cerebrospinal fluid leak, opening of the lateral ventricle during sectioning of the corpus callosum should be avoided. In the parasagittal part, callosal fibers form the upper wall of the lateral ventricle. However, in the midline, callosal sectioning can be achieved without opening the lateral ventricle, which is clarified by opening the cavum septum pellucidum (CSP).

Small vessels on the surface of the corpus callosum are coagulated and divided. Disconnection of the corpus callosum is initiated using microscissors, forceps, a suction tube, or an ultrasonic aspirator. With meticulous dissection, the surgeon enters the CSP without opening the lateral ventricle. Even when the CSP is not clearly visible on preoperative coronal MRI, it can be identified intraoperatively in all cases. Dissection of the corpus callosum is then continued anteriorly along the CSP to disconnect the genu and rostrum of the corpus callosum (Fig. 6.7).

Thereafter, the microscope is redirected posteriorly to dissect the posterior part of the corpus callosum (Fig. 6.8). In anterior corpus callosotomy (aCC), the transitional portion between the body and isthmus represents the posterior limit of disconnection, which can be confirmed using neuronavigation. In total corpus callosotomy (tCC), disconnection is continued posteriorly to the splenium. At this stage, the superior wall of the cavum velum interpositum (CVI), rather than the CSP, serves as a landmark for completing the disconnection of the callosal fibers. After disconnection of the splenium, the vein of Galen is visualized through the arachnoid membrane.

6.1.4 Closure

After confirming hemostasis, the dura is closed, the bone flap is replaced, and the procedure is completed.

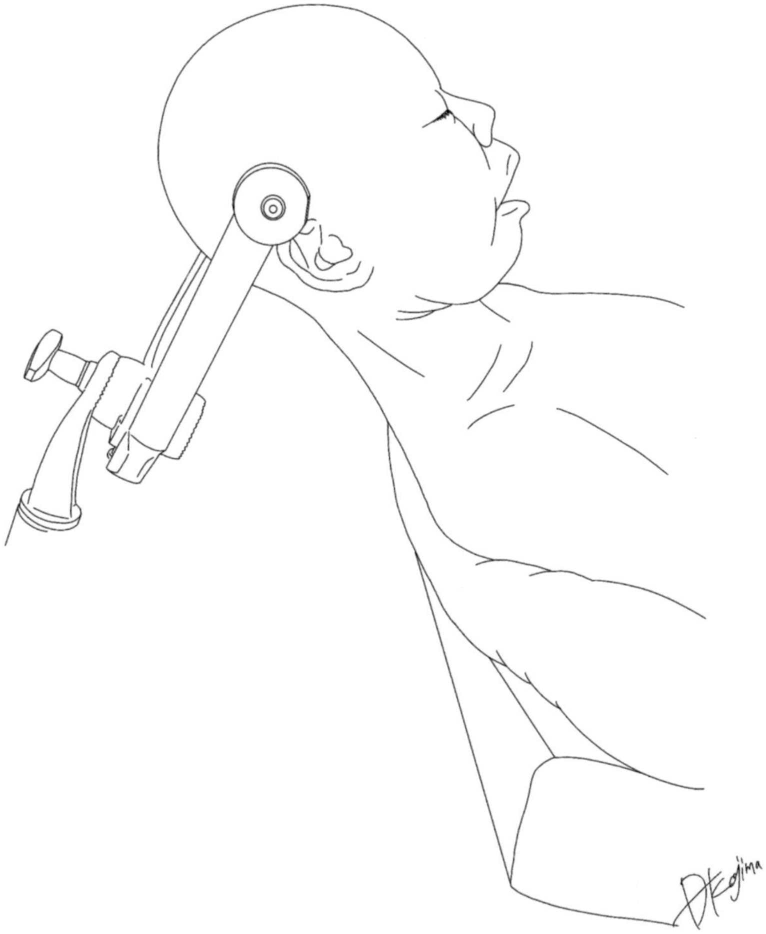

Fig. 6.2 Patient position for total/anterior corpus callosotomy. The patient is placed in the supine position without head rotation

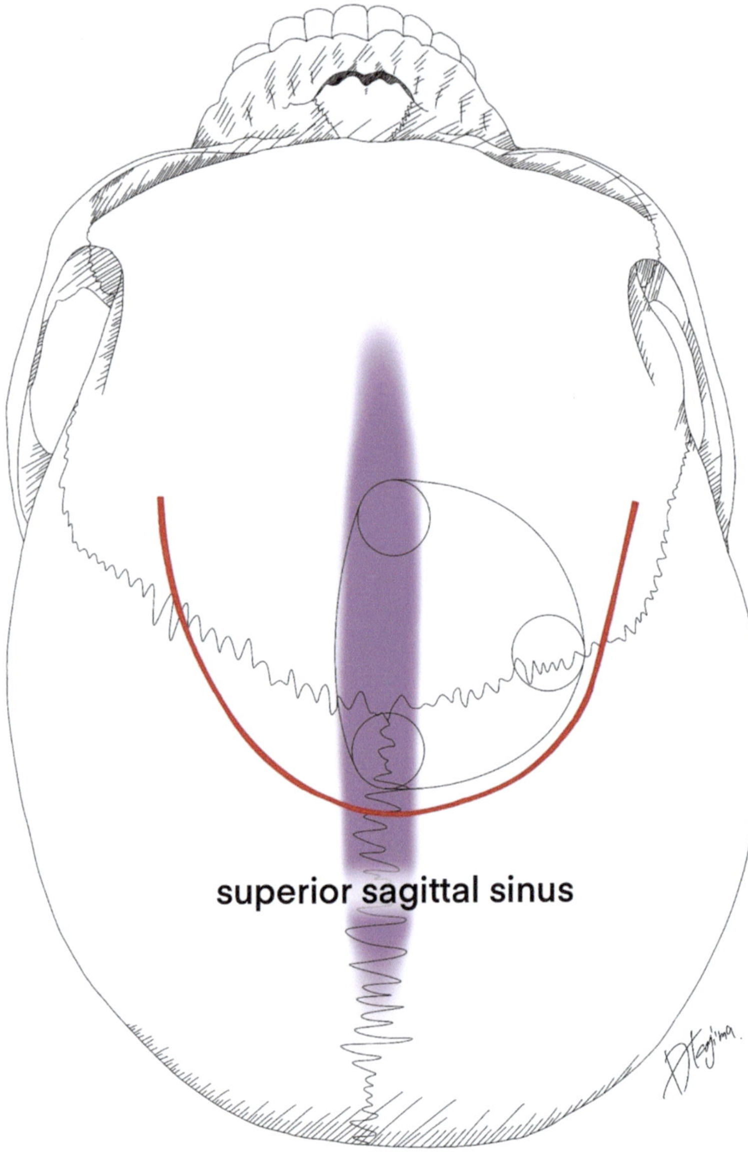

Fig. 6.3 Skin incision and craniotomy for total/anterior corpus callosotomy. A coronal skin incision is used, and the craniotomy is approximately 6 cm in diameter

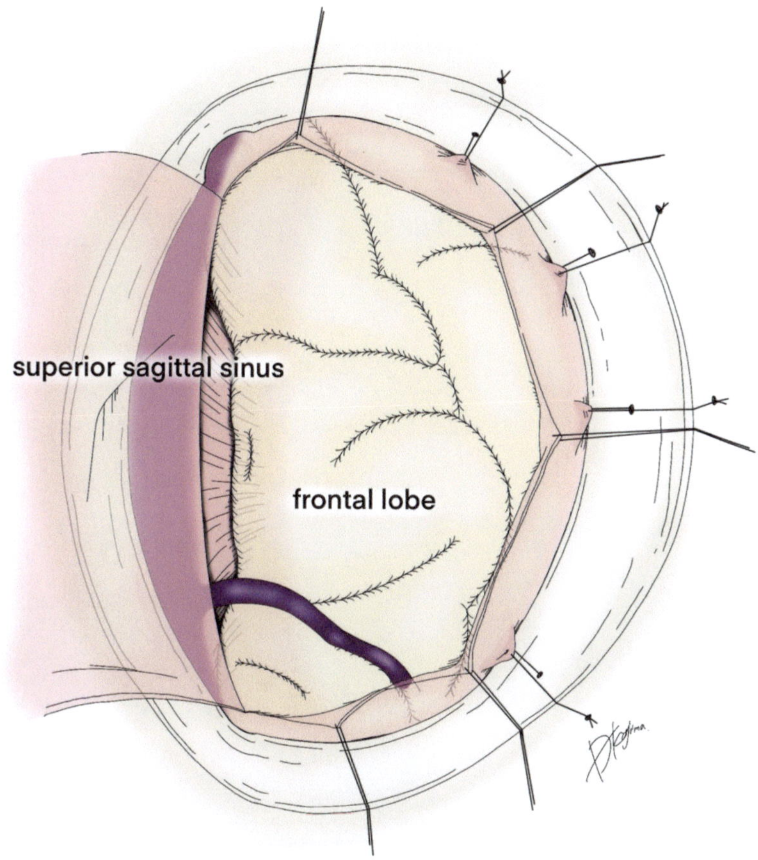

Fig. 6.4 Dural incision and exposure of the frontal lobe. The side of entry into the interhemispheric fissure is determined by the bridging veins to the superior sagittal sinus, favoring the side with fewer obstructing veins. Preoperative venous imaging is useful for identifying bridging veins to the SSS. Dura is opened toward midline

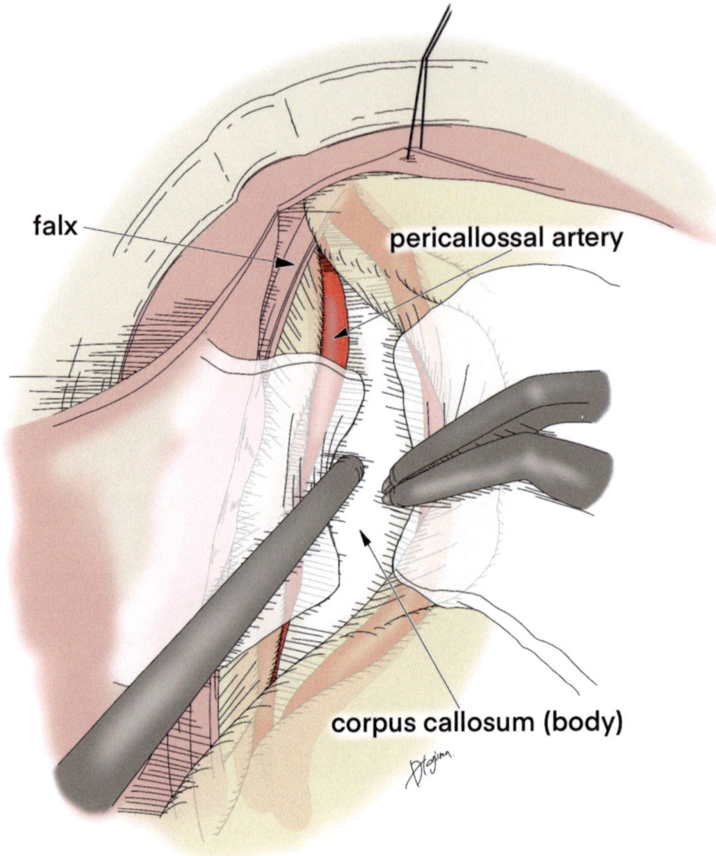

Fig. 6.5 Interhemispheric fissure dissection. After the meticulous dissection of the interhemispheric fissure, the callosomarginal arteries and the pericallosal arteries are identified. Between the bilateral pericallosal arteries, the body of the corpus callosum is seen

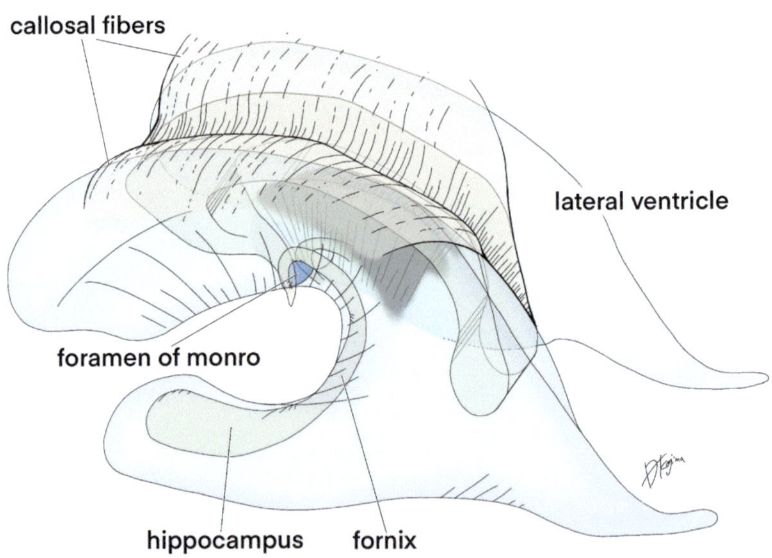

Fig. 6.6 Relationships of the corpus callosum, fornix, and lateral ventricle. The three-dimensional anatomical relationships among the corpus callosum, fornix, and lateral ventricle are demonstrated

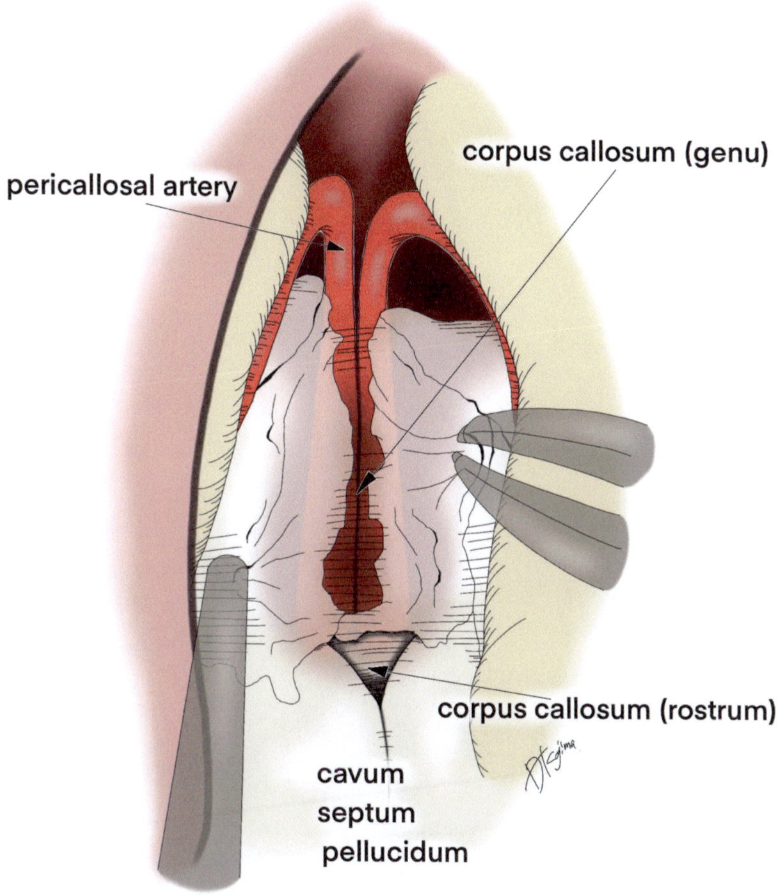

Fig. 6.7 Anterior part of corpus callosotomy. After opening the cavum septum pellucidum (CSP), dissection of the corpus callosum is continued anteriorly along the CSP to disconnect the genu and rostrum of the corpus callosum (Fig. 6.7)

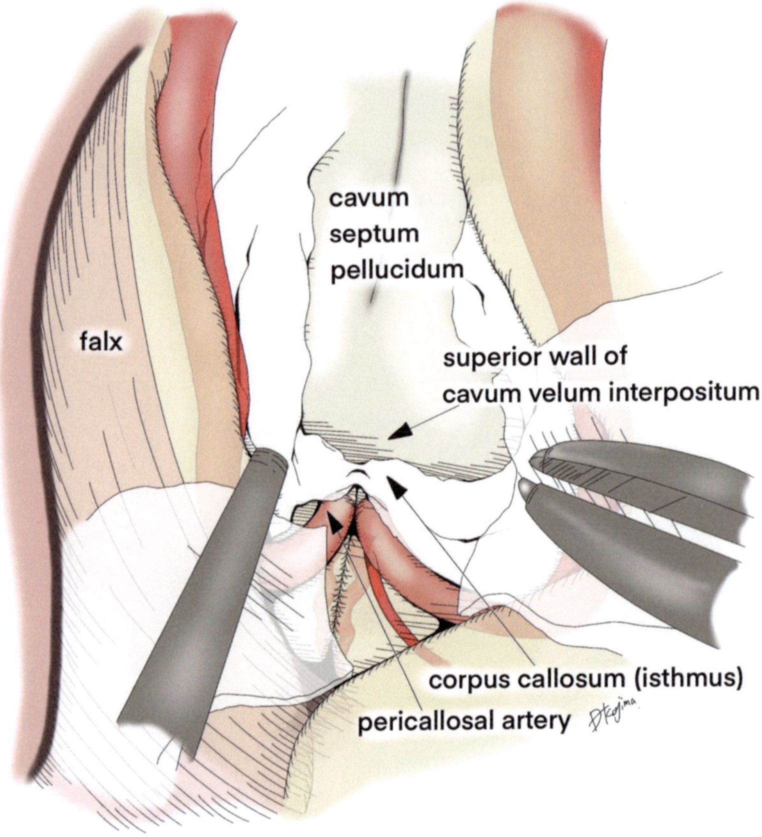

Fig. 6.8 Posterior part of corpus callosotomy. Callosal sectioning continued posteriorly to the isthmus and splenium. At this stage, the superior wall of the cavum velum interpositum (CVI) serves as a landmark for completing disconnection

6.2 Surgical Procedure of Posterior Corpus Callosotomy

6.2.1 Position, Skin Incision, and Craniotomy

When posterior corpus callosotomy (pCC) is performed after anterior corpus callosotomy (aCC), the anterior route to the posterior corpus callosum often has dense adhesions. Therefore, a different and shorter route is selected via a posterior craniotomy.

In this section, we describe the microsurgical technique. However, we currently consider pCC to be a good indication for an endoscope-only approach because at this level, the corpus callosum lies immediately beneath the falx cerebri, eliminating the need for interhemispheric dissection between the bilateral cingulate gyri, and bridging veins are less frequently encountered in this posterior operative corridor. The endoscopic technique is described in a separate section (subsection 6.3.1).

The patient is placed either in a semi-prone park-bench position with the entry side down or in the prone position (Fig. 6.9). A parieto-occipital craniotomy is performed around the lambdoid suture, extending approximately 1 cm beyond the midline. The craniotomy is approximately 5 cm in diameter. A U-shaped or linear skin incision is used (Fig. 6.10).

6.2.2 Interhemispheric Fissure Dissection

Dura is opened toward the midline, and the parietal lobe is retracted laterally (Fig. 6.11). Neuronavigation is useful for selecting the short-

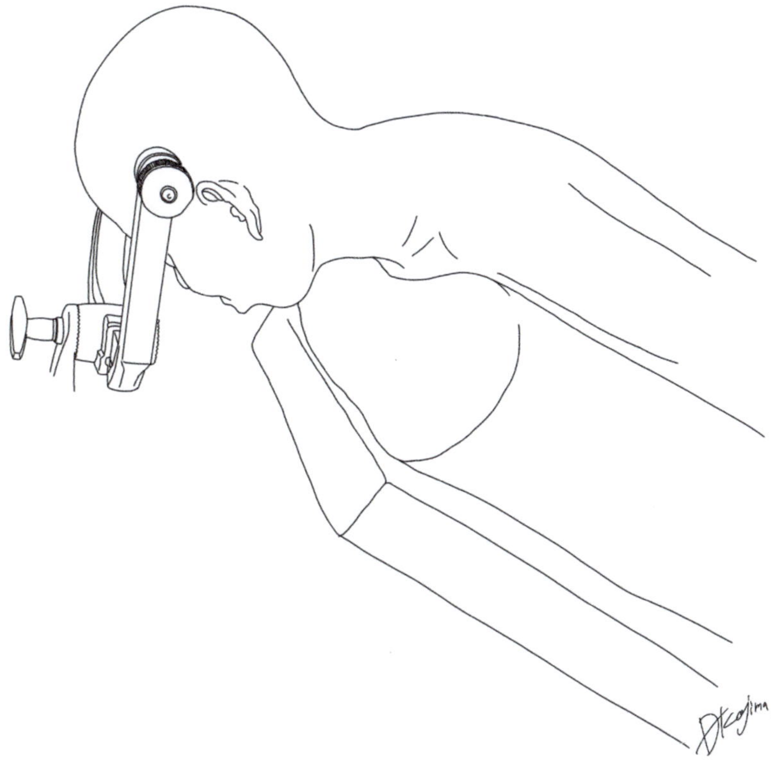

Fig. 6.9 Patient position of posterior one third corpus callosotomy. The patient is placed either in a semi-prone park-bench position with the entry side down or in the prone position

est route to the splenium of the corpus callosum. Drainage of cerebrospinal fluid (CSF) from the pericallosal cistern facilitates brain relaxation. At this level, the inferior edge of the falx cerebri lies close to the corpus callosum, making interhemispheric dissection easier than in the anterior approach. The posterior corpus callosum is identified just below the inferior edge of the falx. Distal segments of the pericallosal arteries may be observed on the corpus callosum.

6.2.3 Severing the Corpus Callosum

The posterior part of the corpus callosum is dissected using microscissors, forceps, a suction tube, or an ultrasonic aspirator. At the posterior margin of the splenium, the vein of Galen is visualized through the arachnoid membrane. Inferiorly, the superior wall of the cavum velum interpositum (CVI) is identified (Fig. 6.12). The microscope is then redirected anteriorly to dissect the isthmus of the corpus callosum (Fig. 6.13).

6.2.4 Closure

After confirming hemostasis, the dura is closed, the bone flap is replaced, and the wound is closed in layers.

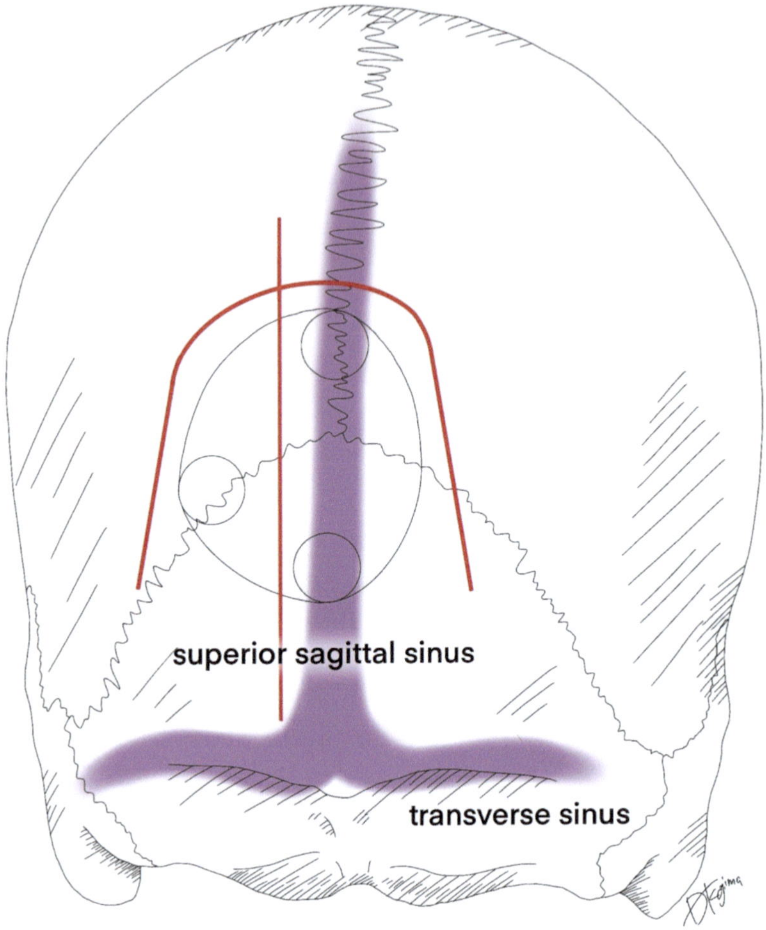

Fig. 6.10 Skin incision and craniotomy for posterior one-third corpus callosotomy. A parieto-occipital craniotomy is performed around the lambdoid suture, extending approximately 1 cm beyond the midline. The craniotomy is approximately 5 cm in diameter. A U-shaped or linear skin incision is used

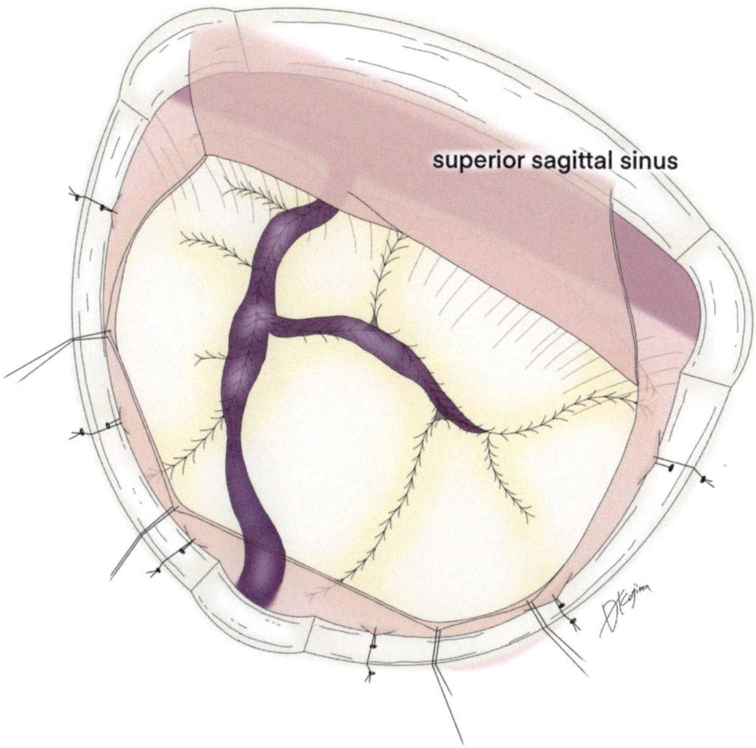

Fig. 6.11 Craniotomy and dural incision for posterior one corpus callosotomy. Dura is opened toward midline. Direction and distance to the splenium is confirmed by neuronavigation

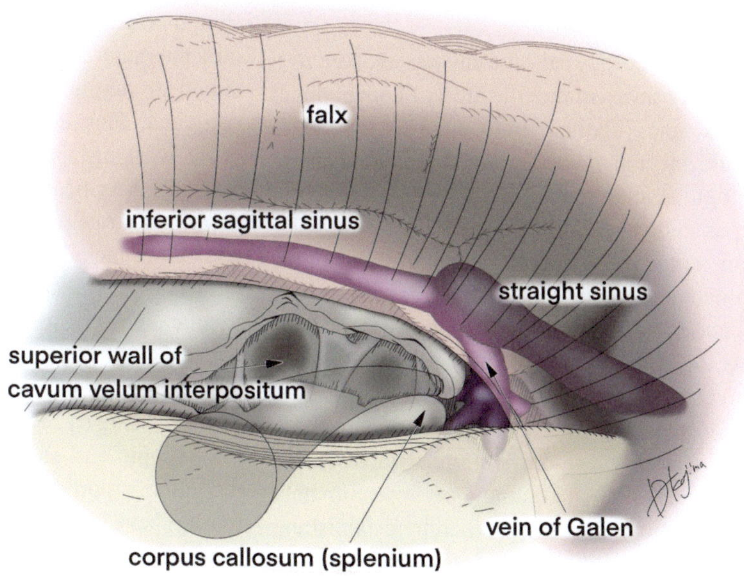

Fig. 6.12 Severing the posterior part of the corpus callosum. The posterior part of the corpus callosum is dissected. At the posterior margin of the splenium, the vein of Galen is visualized through the arachnoid membrane. Inferiorly, the superior wall of the cavum velum interpositum (CVI) is identified

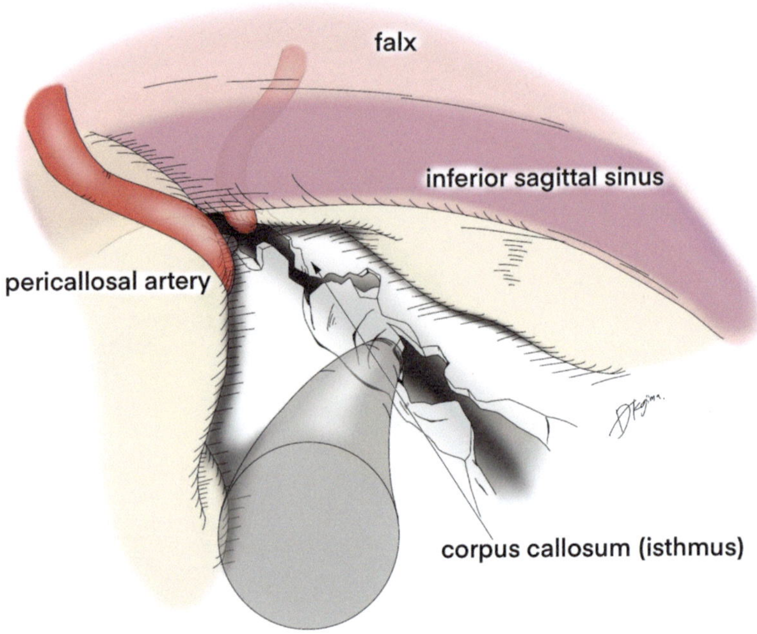

Fig. 6.13 Disconnection of the isthmus of corpus callosotomy. After sectioning the splenium, the microscope is redirected anteriorly and dissection of the isthmus of the corpus callosum is performed

6.3 Note

6.3.1 Endoscopic Corpus Callosotomy

Endoscopic corpus callosotomy has evolved as a minimally invasive modification of conventional microscopic corpus callosotomy. In 2012, Bahuleyan et al. demonstrated through cadaveric anatomical studies that the entire length of the corpus callosum, from the rostrum to the splenium, could be accessed using an endoscopic approach, providing the first clear anatomical evidence that a purely endoscopic corpus callosotomy is feasible [1]. In 2015, Sood et al. reported clinical cases in which corpus callosotomy was completed using an endoscope as the primary visual modality without a surgical microscope, through a small craniotomy of approximately 2–3 cm (microcraniotomy), thereby establishing the clinical feasibility of endoscopic corpus callosotomy [2].

Smyth et al. compared conventional microscopic craniotomy with endoscopic techniques and noted that endoscopic surgery may offer advantages in visualization within narrow surgical corridors and reduced brain retraction while also acknowledging limitations related to hemostasis and instrument maneuverability. They concluded that endoscopic corpus callosotomy should be regarded not as a new standard procedure but rather as one of several technical options aimed at minimizing surgical invasiveness. In particular, they emphasized that securing adequate working space within a limited surgical corridor is critical in endoscopic surgery and described strategies to achieve sufficient brain relaxation, including the administration of mannitol and cerebrospinal fluid drainage via ventricular catheterization when necessary [3].

At our institution, both total corpus callosotomy and anterior corpus callosotomy are performed using a hybrid technique combining microscopic and endoscopic approaches [4]. In the anterior portion, the bilateral cingulate gyri located beneath the falx cerebri are often tightly adherent; therefore, dissection of the interhemispheric fissure is performed under microscopic visualization, while deep corpus callosal sectioning is conducted using an endoscope. In contrast, posterior corpus callosotomy is primarily performed using a purely endoscopic technique [5].

This approach is based on the anatomical fact that the corpus callosum lies directly beneath the falx cerebri in this region, eliminating the need for extensive fissure dissection, and that obstructive bridging veins are encountered less frequently in the posterior interhemispheric corridor.

For endoscopic posterior corpus callosotomy, preoperative assessment focuses on identifying venous structures that may interfere with the surgical corridor. The patient is positioned prone, and neuronavigation is used to determine the posterior margin of the corpus callosum and the posterior border of the previous anterior callosotomy. A small craniotomy of approximately 3 cm is planned to allow the shortest trajectory to these targets. When no obstructing veins are present, we generally approach from the left side. After craniotomy and dural opening, a rigid endoscope is advanced into the interhemispheric fissure to reach the corpus callosum. Brain relaxation facilitates manipulation within the fissure. Sectioning begins at the splenium, and once the splenium is divided, the vein of Galen becomes visible. After confirming complete posterior disconnection, the dissection proceeds anteriorly until it is connected with the posterior margin of the previously performed anterior corpus callosotomy.

6.3.2 Outcomes After Corpus Callosotomy

Chan et al. conducted a systematic review and meta-analysis of 58 studies comprising 1742 adult and pediatric patients to evaluate seizure outcomes after corpus callosotomy. At final follow-up, the rate of complete seizure freedom was 18.8%, while the disappearance rate of drop attacks was 55.3%, indicating that corpus callosotomy is particularly effective in controlling drop attacks. Predictors of complete seizure freedom included a history of infantile spasms, normal preoperative MRI findings, and shorter duration. Disappearance of drop attacks was more frequent after total corpus callosotomy than after partial callosotomy and was more likely in patients with idiopathic epilepsy. Perioperative adverse events were reported in approximately 8–12% of cases, with transient lower limb weakness, aphasia or mutism, and infection being the most common. Importantly, no consistent evidence demonstrated a significantly higher incidence of disconnection syndrome following total corpus callosotomy [6]. Wu et al. analyzed 49 studies involving 1644 patients with a minimum follow-up of one year to evaluate long-term seizure outcomes after corpus callosotomy for drug-resistant epilepsy. While the complete seizure freedom rate was limited to 12.4%, the disappearance rate of drop attacks reached 61.9%. In procedure-specific analyses, total corpus callosotomy was associated with significantly higher rates of drop attack elimination than anterior corpus callosotomy, whereas no significant difference was observed between the two procedures in terms of complete seizure freedom [7]. Whitney et al. investigated the efficacy of corpus callosotomy in adult and pediatric patients with drug-resistant epileptic spasms (ES) and examined clinical factors influencing seizure outcomes. This meta-analysis included 12 studies and 253 patients, with a mean follow-up of 47 months. The spasm freedom rate was 31% (95% CI: 22–42%). Structural etiology and older age at surgery were identified as unfavorable prognostic factors. Approximately one-quarter of patients underwent additional epilepsy surgery for residual ES, and more than half of these achieved eventual spasm freedom [8]. These findings from meta-analyses are supported by relatively large single-center retrospective studies. Aum et al. evaluated 103 patients under 21 years of age who underwent corpus callosotomy between 2003 and 2021 and found that 75% of patients with preoperative drop attacks achieved complete elimination of these seizures. The overall surgical complication rate was low at 6%. No significant difference in overall seizure outcomes was observed between anterior and total corpus callosotomy, and laser interstitial thermal therapy (LITT) demonstrated seizure control comparable to open surgery. Furthermore, long-term follow-up studies indicate that recurrence of drop attacks most commonly occurs within 12 months postoperatively, and patients who achieve favorable early outcomes are more likely to maintain sustained seizure control [9]. In summary, Chan et al. comprehensively characterized seizure outcomes and prognostic factors following corpus callosotomy, Wu et al. compared long-term outcomes between total and anterior corpus callosotomy, and Whitney et al. demonstrated that etiology and timing of surgery are critical determinants of outcome in patients with epileptic spasms. Corpus callosotomy should be selected as a palliative surgical procedure based on seizure type, and its efficacy is strongly influenced by patient characteristics and timing of intervention. The most characteristic complication following corpus callosotomy is disconnection syndrome, which is reported to be transient in most cases [10]. Clinically, this syndrome manifests as reduced speech output, impaired bimanual coordination, and gait ataxia. These symptoms are more common after total corpus callosotomy than after anterior corpus callosotomy [11, 12]. Previous studies have reported transient disconnection syndrome in approximately 10–20% of patients undergoing total corpus callosotomy [10], and permanent severe neurological deficits are considered rare [10, 13].

References

1. Bahuleyan B, Vogel TW, Robinson S, Cohen AR. Endoscopic total corpus callosotomy: cadaveric demonstration of a new approach. Pediatr Neurosurg. 2011;47(6):455–60. https://doi.org/10.1159/000338984.
2. Sood S, Marupudi NI, Asano E, Haridas A, Ham SD. Endoscopic corpus callosotomy and hemispherotomy. J Neurosurg Pediatr. 2015;16(6):681–6. https://doi.org/10.3171/2015.5.PEDS1531.
3. Smyth MD, Vellimana AK, Asano E, Sood S. Corpus callosotomy—open and endoscopic surgical techniques. Epilepsia. 2017;58(Suppl 1):73–9. https://doi.org/10.1111/epi.13681.
4. Yindeedej V, Uda T, Kawashima T, Koh S, Tanoue Y, Kojima Y, et al. Use of an endoscope reduces the size of craniotomy without increasing operative time compared with conventional microscopic corpus callosotomy. Oper Neurosurg. 2023;25(6):505–11. https://doi.org/10.1227/ons.0000000000000867.
5. Uda T, Kunihiro N, Umaba R, Koh S, Kawashima T, Ikeda S, et al. Surgical aspects of corpus callosotomy. Brain Sci. 2021;11(12):1608. https://doi.org/10.3390/brainsci11121608.
6. Chan AY, Rolston JD, Lee B, Vadera S, Englot DJ. Rates and predictors of seizure outcome after corpus callosotomy for drug-resistant epilepsy: a meta-analysis. J Neurosurg. 2018;130(4):1193–202. https://doi.org/10.3171/2017.12.JNS172331.
7. Wu X, Ou S, Zhang H, Zhen Y, Huang Y, Wei P, et al. Long-term follow-up seizure outcomes after corpus callosotomy: a systematic review with meta-analysis. Brain Behav. 2023;13(4):e2964. https://doi.org/10.1002/brb3.2964.
8. Whitney R, Otsubo H, Cunningham J, Jones KC, Ramachandran Nair R, Nouri MN, et al. Corpus callosotomy for refractory epileptic spasms: a systematic review and meta-analysis. Seizure. 2024;123:159–67. https://doi.org/10.1016/j.seizure.2024.11.001.
9. Aum DJ, Reynolds RA, McEvoy S, Tomko S, Zempel J, Roland JL, et al. Surgical outcomes of open and laser interstitial thermal therapy approaches for corpus callosotomy in pediatric epilepsy. Epilepsia. 2023;64(9):2274–85. https://doi.org/10.1111/epi.17679.
10. Graham D, Tisdall MM, Gill D. Corpus callosotomy outcomes in pediatric patients: a systematic review. Epilepsia. 2016;57(7):1053–68. https://doi.org/10.1111/epi.13408.
11. Cendes F, Ragazzo PC, da Costa V, Martins LF. Corpus callosotomy in treatment of medically resistant epilepsy: preliminary results in a pediatric population. Epilepsia. 1993;34(5):910–7. https://doi.org/10.1111/j.1528-1157.1993.tb02111.x.
12. Reutens DC, Bye AM, Hopkins IJ, Danks A, Somerville E, Walsh J, et al. Corpus callosotomy for intractable epilepsy: seizure outcome and prognostic factors. Epilepsia. 1993;34(5):904–9. https://doi.org/10.1111/j.1528-1157.1993.tb02110.x.
13. Graham D, Gill D, Dale RC, Tisdall MM. Corpus Callosotomy outcomes study group. Seizure outcome after corpus callosotomy in a large paediatric series. Dev Med Child Neurol. 2018;60(2):199–206. https://doi.org/10.1111/dmcn.13592.

7 Frontal Lobe Disconnection (FD)

7.1 Surgical Procedures

7.1.1 Patient Positioning and Head Fixation

The patient is placed in a supine-lateral position. The trunk is slightly tilted laterally, and the head is rotated markedly toward the contralateral side of the lesion. This positioning allows the lateral surface of the frontal lobe on the operative side to be viewed almost directly from a lateral trajectory.

Neuronavigation is used to approximate the location of the central sulcus and the Sylvian fissure, allowing confirmation that these key landmarks are adequately covered by the planned craniotomy.

7.1.2 Skin Incision and Craniotomy

In frontal lobe disconnection, one side of the frontal lobe is disconnected from the rest part of the brain, preserving the precentral gyrus. Then the precentral sulcus serves as a posterior limit of disconnection (Fig. 7.1). A curvilinear skin incision is made beginning anterior to the auricle and extending posteriorly in a semicircular fashion, crossing the midline as necessary to ensure adequate exposure of the central sulcus. The scalp and temporalis muscle are reflected together as a single myocutaneous flap.

Using neuronavigation, the precentral sulcus—serving as a reference for the intended disconnection plane—is precisely identified. The craniotomy is designed to provide sufficient exposure of the precentral sulcus. Superiorly, the exposure extends to the midline over the superior sagittal sinus, and inferiorly, it includes the region of the Sylvian fissure.

Burr holes are placed at the frontozygomatic suture, along the squamous suture, at the midline over the superior sagittal sinus, and near the coronal suture. These burr holes are connected to create a wide fronto-parieto-temporal craniotomy.

To allow broad visualization of the frontal cortical margin corresponding to the disconnection plane, the dura is opened in a radial or H-shaped fashion, providing generous operative exposure (Fig. 7.2).

7.1.3 Identification of the Central Sulcus and Posterior Frontal Disconnection

Somatosensory evoked potentials (SEPs) are first recorded using median nerve stimulation. The central sulcus is identified by confirming phase reversal of the cortical responses. Cortical motor evoked potentials (MEPs) are subsequently obtained to functionally localize the primary motor cortex. The combined use of these neuro-

T. Uda et al., *Atlas of Epilepsy Surgery*, https://doi.org/10.1007/978-981-92-0176-1_7

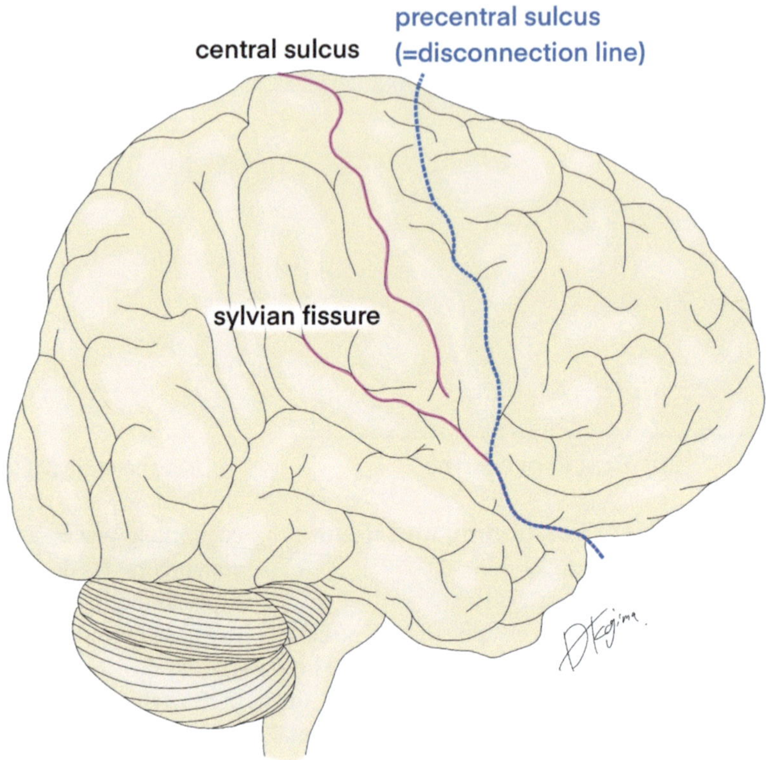

Fig. 7.1 Schematic illustration of frontal lobe disconnection. In frontal lobe disconnection, one side of the frontal lobe is disconnected from the rest of the brain, preserving the precentral gyrus

physiological monitoring techniques enables the precise identification of the precentral sulcus, which defines the posterior limit of the disconnection plane.

A corticotomy is then performed in the superior frontal gyrus anterior to the precentral sulcus, and the dissection proceeds into the subcortical white matter. Disconnection is first advanced medially toward the falx cerebri until the interhemispheric fissure is reached. After completing the medial disconnection, the resection plane is extended laterally.

The corticotomy is then extended from the middle frontal gyrus to the inferior frontal gyrus. The disconnection proceeds deeper while dividing the frontal operculum, reaching a depth corresponding to the level of the insular cortex. Then, the body of the lateral ventricle is opened (Fig. 7.3).

7.1.4 Disconnection of the Lateral and Basal Frontal Lobe

While maintaining the depth at the insular cortex, the disconnection is advanced anteriorly. The dissection is extended toward the region connecting the anterior insular cortex and the sphenoid ridge, ultimately reaching the frontal base (Fig. 7.4).

The sphenoid ridge is then followed medially, and the basal frontal cortex—including the orbital gyri and the gyrus rectus—is disconnected. During disconnection along the anterior skull base, the olfactory nerve is identified. The dissection proceeds further medially until the interhemispheric fissure is reached (Fig. 7.5).

At this stage, the proximal A2 segment of the anterior cerebral artery (pericallosal artery), serving as a landmark for the completion of the medial disconnection, is identified.

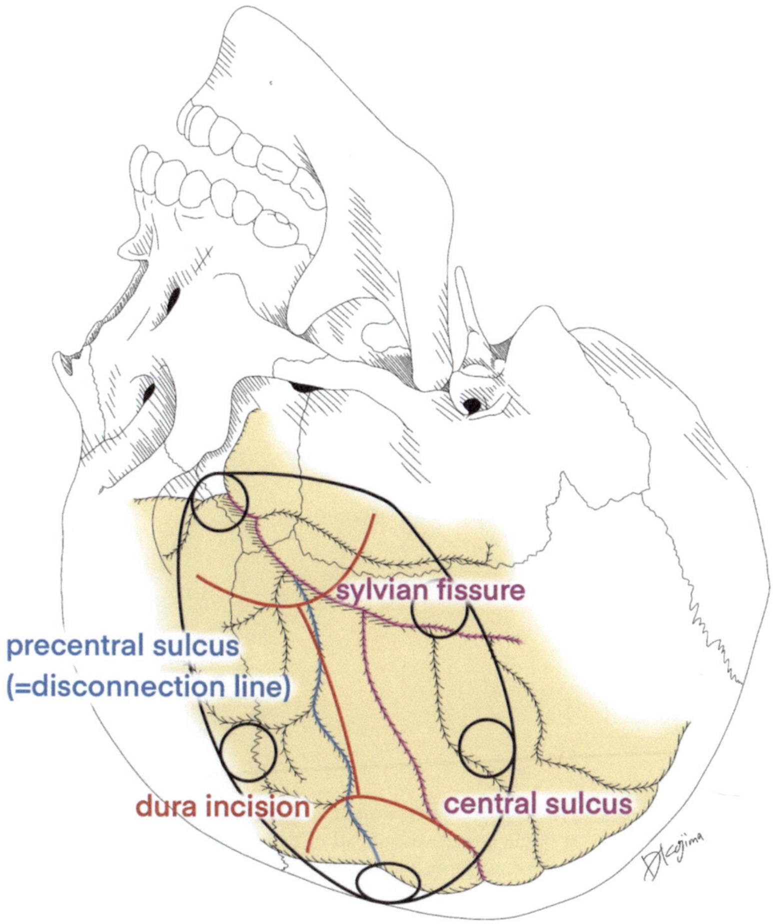

Fig. 7.2 Craniotomy and dural incision for frontal lobe disconnection. A wide fronto-temporo-parietal craniotomy was designed to provide sufficient exposure of the precentral sulcus. Burr holes are placed at the frontozygomatic suture, along the squamous suture, at the midline over the superior sagittal sinus, and near the coronal suture. To allow broad visualization of the frontal cortical margin corresponding to the disconnection plane, the dura is opened in a radial fashion

7.1.5 Disconnection of the Inferior and Medial Frontal Lobe

From the lateral ventricle, the disconnection plane is extended medially along the ventricular wall under subpial dissection toward the falx cerebri. Using the falx as a medial landmark, dissection proceeds inferiorly along the interhemispheric surface. The cingulate gyrus is removed up to the midline, exposing the pericallosal artery and the corpus callosum (Fig. 7.6). Using the pericallosal artery as a guide, the corpus callosum is divided toward its proximal portion. The disconnection line is then connected to the previously established anterior disconnection plane (Fig. 7.7).

7.1.6 Closure

After confirming hemostasis, the dura is closed with sutures, and the bone flap is replaced to complete the procedure.

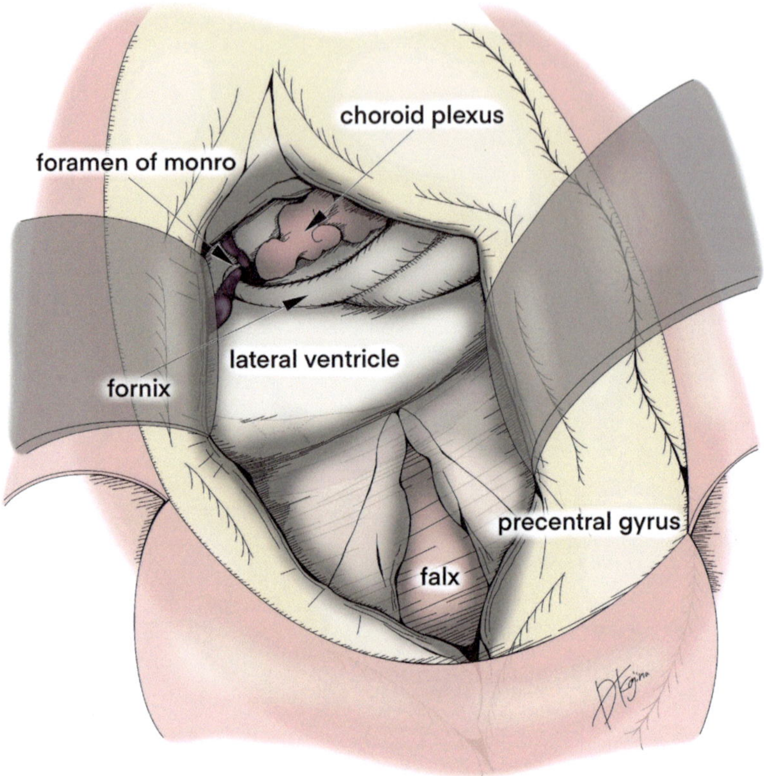

Fig. 7.3 Disconnection of the postero-lateral part of the frontal lobe. A corticotomy is performed in the superior frontal gyrus anterior to the precentral sulcus, and the dissection proceeds into the subcortical white matter. Disconnection is first advanced medially toward the falx cerebri. After completing the medial disconnection, the resection plane is extended laterally. The corticotomy is then extended from the middle frontal gyrus to the inferior frontal gyrus. The disconnection proceeds deeper while dividing the frontal operculum, reaching a depth corresponding to the level of the insular cortex. Then the body of lateral ventricle is opened

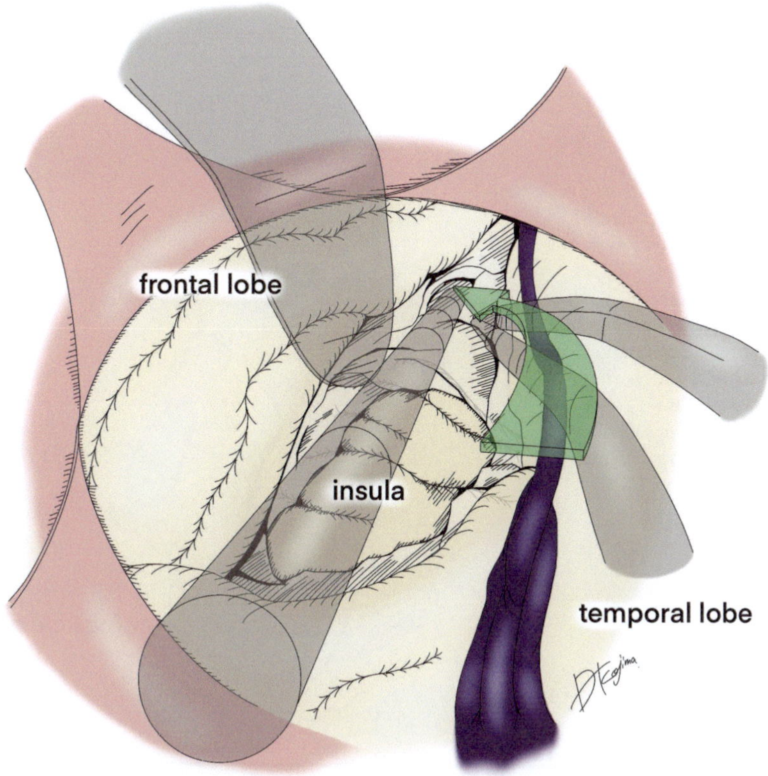

Fig. 7.4 Disconnection of the lateral part of the frontal lobe. Maintaining the depth at the level of the insular cortex, the disconnection is advanced anteriorly. The dissection is extended toward the region connecting the anterior insular cortex and the sphenoid ridge, ultimately reaching the frontal base

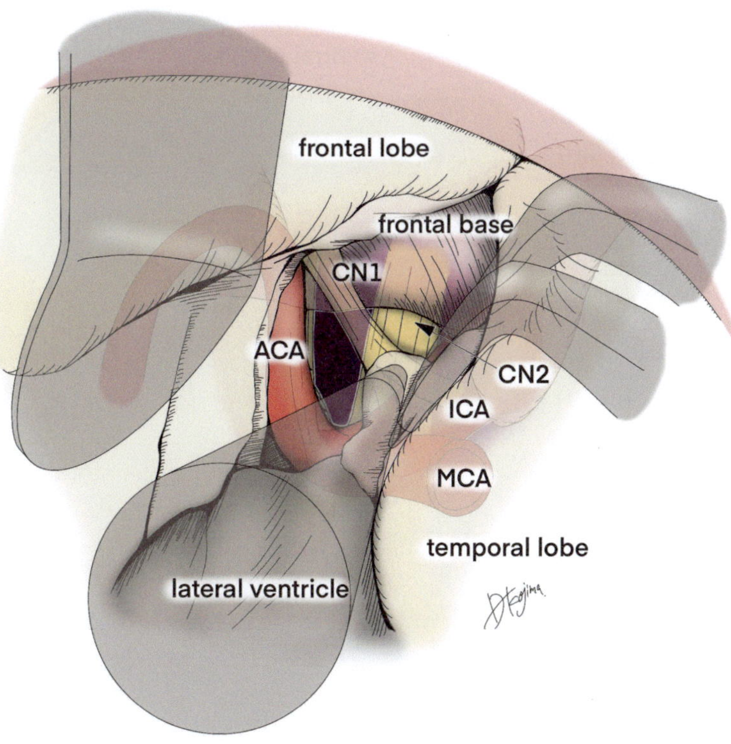

Fig. 7.5 Antero-medial disconnection. Following sphenoid ridge medially, the orbital and rectal gyrus are disconnected. At the olfactory sulcus, the olfactory nerve is identified. The dissection proceeds further medially until the interhemispheric fissure, where the proximal A2 segment of the anterior cerebral artery is identified (*ACA* anterior cerebral artery, *CN1* olfactory nerve, *CN2* optic nerve, *ICA* internal cerebral artery, *MCA* middle cerebral artery)

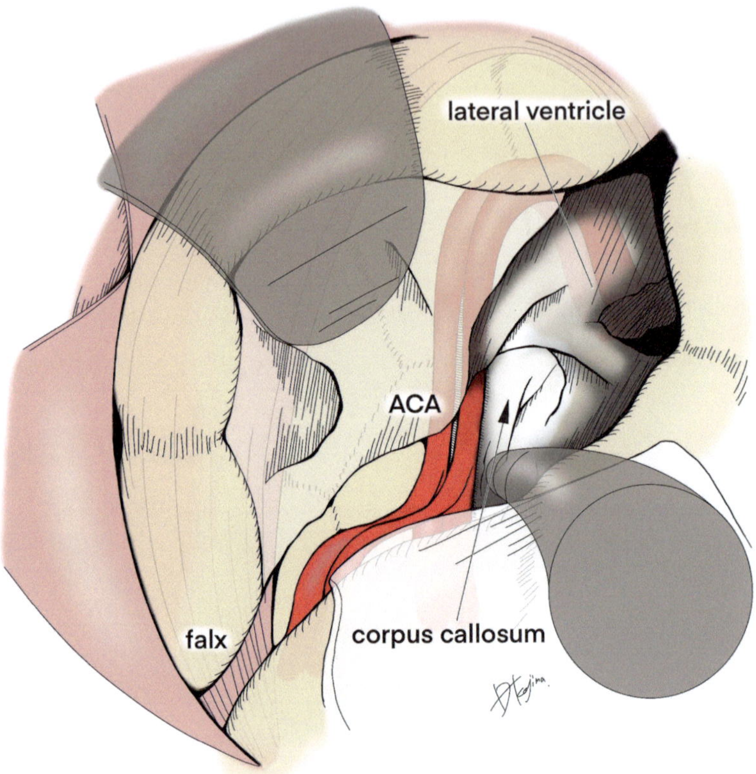

Fig. 7.6 Postero-medial disconnection of the frontal lobe. From the lateral ventricle, the disconnection plane is extended medially along the ventricular wall under subpial dissection toward the falx cerebri. The dissection proceeds inferiorly along the interhemispheric surface. The cingulate gyrus is removed up to the midline, exposing the pericallosal artery and the corpus callosum (*ACA* anterior cerebral artery)

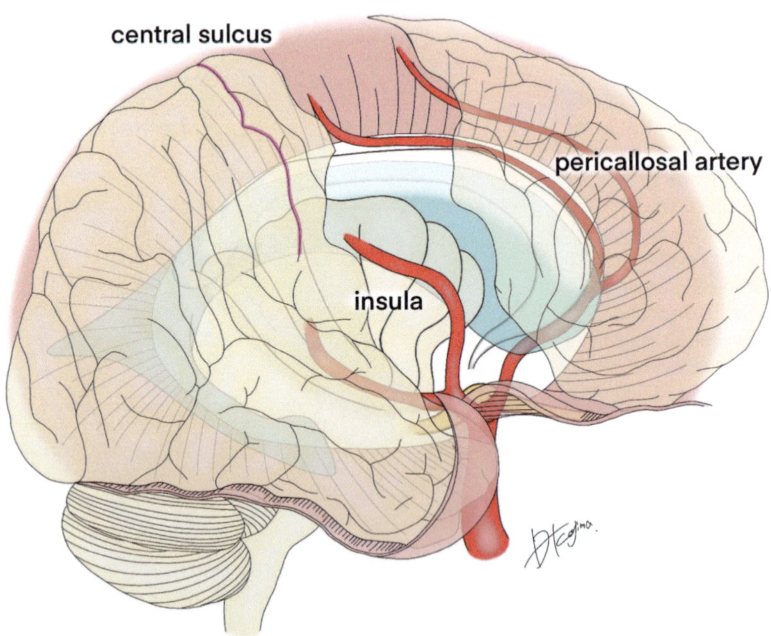

Fig. 7.7 Achievement of the frontal lobe disconnection. Using the pericallosal artery as a guide, the corpus callosum is divided toward its proximal portion. The disconnection is connected to the previously established anterior disconnection plane

7.2 Notes

7.2.1 Disconnection Strategy According to Hemispheric Dominance

The surgical technique described in Sect. 9.1 is primarily intended for procedures involving the nondominant hemisphere. In contrast, frontal lobe disconnection (FD) in the language-dominant hemisphere requires special consideration, as language function is frequently localized in the posterior portion of the inferior frontal gyrus. Therefore, careful attention must be paid when determining the posterior limit of the disconnection plane in this region.

Accordingly, the posterior boundary of disconnection in the dominant hemisphere should be tailored to the individual functional anatomy of each patient.

To preserve language function, intraoperative language mapping under awake conditions may be considered, if feasible. When awake mapping is not practical, functional evaluation using cortico-cortical evoked potentials (CCEP) may provide an alternative approach.

The use of these techniques may help increase the likelihood of functional preservation while maintaining adequate disconnection in the dominant hemisphere.

7.2.2 Resection and Disconnection in Frontal Epilepsy Surgery

Frontal lobectomy (FL) has traditionally been widely performed as a focal surgical treatment for frontal lobe epilepsy, and substantial data regarding its surgical outcomes have accumulated.

Regarding the surgical outcomes of FL, Mosewich et al. reported a retrospective analysis of 68 patients, in which approximately 59% achieved seizure freedom or experienced only minor seizures not interfering with daily life at final follow-up. Outcomes were more favorable in patients with magnetic resonance imaging (MRI)-localized frontal lesions: approximately 72% of lesional cases achieved good outcomes, compared with 41% of nonlesional cases. Major complications were observed in approximately 4% of patients, including mild to moderate hemiparesis and aphasia [1].

In contrast, Jeha et al. performed a longitudinal analysis of seizure outcomes in 70 patients who underwent FL and demonstrated that complete seizure freedom declined over time, with rates of approximately 56% at 1 year, 45% at 3 years, and 30% at 5 years postoperatively. Approximately 80% of recurrences occurred within 6 months after surgery. Notably, among patients who underwent complete resection, approximately 66% maintained seizure freedom at 3 years, whereas only 11% of those with incomplete resection remained seizure-free, underscoring the importance of achieving complete resection. Independent predictors of poor outcome included MRI-negative cortical dysplasia, extrafrontal MRI abnormalities, generalized or nonlocalizing ictal electroencephalogram (EEG) patterns, acute postoperative seizures (APOS), and incomplete resection [2].

In recent years, reports on frontal lobe disconnection (FD) have been increasing [3, 4]. Kamalboor et al. described 16 patients who underwent FD for drug-resistant frontal lobe epilepsy, of whom eight patients (50%) achieved seizure freedom. Although the number of cases was limited, serious surgery-related complications were observed in three patients (18.8%), including right subgaleal and epidural abscess, frontal epidural hematoma, and intraoperative hemorrhage. Neurological deficits were also reported in three patients (18.8%), primarily presenting as mild contralateral limb weakness due to the involvement of the supplementary motor area or primary motor cortex. In addition, one patient experienced a mild decline in auditory verbal learning and memory as well as visuospatial cognitive function [3].

At present, no direct comparative studies between FL and FD are available. However, in hemisphere surgery, functional hemispherotomy has been reported to be associated with fewer postoperative complications compared with anatomical hemispherectomy [5], leading to a historical shift from resection-based to disconnection-based procedures. Given this background, a similar transition from resection to disconnection in frontal lobe surgery may therefore be reasonable.

7.2.3 Surgical Considerations for the Frontal Operculum

Focal surgical approaches for frontal lobe epilepsy, including FL and FD, have been described above. These techniques encompass most regions of the frontal lobe, except for the precentral gyrus. However, the frontal operculum is a region that may remain outside the typical extent of resection or disconnection.

Regarding focal resection of the frontal operculum, d'Orio et al. analyzed 65 patients who underwent surgery for suprasylvian operculoinsular epilepsy and reported that 62.9% achieved Engel class I at final follow-up (median 56 months). Immediate postoperative neurological deficits were observed in 46.2% of patients; however, most were transient, and permanent deficits persisting beyond 6 months were present in 12.3%. Postoperative MRI demonstrated ischemic lesions of the corona radiata in 67.7% of cases, and corticospinal tract Wallerian degeneration (CSTWd) was identified in 29.2%, which was independently associated with permanent neurological deficits. Combined operculoinsular resection was identified as a risk factor for white matter ischemia, whereas selective opercular resection was associated with a reduced risk of seizure recurrence [6].

These findings suggest that in surgeries involving the opercular region, deep white matter ischemia rather than cortical resection itself may play a critical role in determining neurological outcomes. A similar consideration applies to insular glioma surgery, in which meticulous techniques aimed at avoiding deep white matter ischemia are essential.

References

1. Mosewich RK, So EL, O'Brien TJ, Cascino GD, Sharbrough FW, Marsh WR, et al. Factors predictive of the outcome of frontal lobe epilepsy surgery. Epilepsia. 2000;41(7):843–9. https://doi.org/10.1111/j.1528-1157.2000.tb00251.x.
2. Jeha LE, Najm I, Bingaman W, Dinner D, Widdess-Walsh P, Lüders H. Surgical outcome and prognostic factors of frontal lobe epilepsy surgery. Brain. 2007;130(Pt 2):574–84. https://doi.org/10.1093/brain/awl364.
3. Kamalboor H, Alhindi H, Alotaibi F, Althubaiti I, Alkhateeb M. Frontal disconnection surgery for drug-resistant epilepsy: outcome in a series of 16 patients. Epilepsia Open. 2020;5(3):475–86. https://doi.org/10.1002/epi4.12424.
4. Yindeedej V, Uda T, Kunihiro N, Goto T. Frontal lobe disconnection: how I do it. Acta Neurochir. 2024;166(1):429. https://doi.org/10.1007/s00701-024-06319-0.
5. Delalande O, Bulteau C, Dellatolas G, Fohlen M, Jalin C, Buret V, et al. Vertical parasagittal hemispherotomy: surgical procedures and clinical long-term outcomes in a population of 83 children. Neurosurgery. 2007;60(2 Suppl 1):ONS19–32.; discussion ONS32. https://doi.org/10.1227/01.NEU.0000249246.48299.12.
6. d'Orio P, Squarza S, Revay M, Cardinale F, Castana L, Sartori I, et al. Neurological morbidity of surgery for suprasylvian operculoinsular epilepsy. Epilepsia. 2024;65(2):402–13. https://doi.org/10.1111/epi.17844.

8 Posterior Quadrant Disconnection (PQD)/Temporo-Parieto-Occipital Disconnection

8.1 Surgical Procedures

8.1.1 Patient Positioning and Head Fixation

The patient is placed in the spine-lateral position. The head is rotated approximately 90° toward the contralateral side and fixed with a slight vertex-down tilt. The projected Sylvian fissure, central sulcus, and postcentral sulcus are marked on the scalp in advance.

For neurophysiological monitoring, somatosensory evoked potentials (SEPs) elicited by median nerve stimulation, motor evoked potentials (MEPs), and transcranial motor evoked potentials (TcMEPs) are prepared, although the monitoring modalities used may vary among institutions. In the dominant hemisphere, cortico-cortical evoked potentials (CCEPs) can also be informative.

8.1.2 Skin Incision and Craniotomy

The skin incision is made as a large curvilinear arc extending posteriorly from the preauricular region, allowing sufficient exposure of the postcentral sulcus up to near the midline. Craniotomy is performed as an elliptic parietotemporal craniotomy to provide wide exposure from the proximal to distal Sylvian fissure, as well as along the entire length of the postcentral sulcus (Fig. 8.1). The dura is opened and reflected in a radial fashion to allow visualization of the Sylvian fissure and the entire central sulcus.

8.1.3 Identification of Functional Areas

First, the central sulcus is identified using SEPs. The primary motor cortex is then confirmed by MEPs. Based on these landmarks, the postcentral sulcus is determined accurately. In the language-dominant hemisphere, CCEPs are additionally recorded after craniotomy to evaluate functional connectivity before disconnection.

8.1.4 Sylvian Fissure Dissection and Temporal Disconnection

The Sylvian fissure is widely dissected from anterior to posterior up to the posterior insula. Next, a cortical incision is made along the inferior periinsular sulcus, the temporal stem is transected, and the temporal horn of the lateral ventricle is entered. After reaching the temporal horn, the incision through the temporal stem is extended anteriorly to the temporal pole. In the deep portion, the disconnection is carried down to the pia mater.

A portion of the amygdala facing the hippocampus is resected. Within the temporal horn, the choroid plexus and inferior choroidal point are identified as anatomical landmarks. From this

T. Uda et al., *Atlas of Epilepsy Surgery*, https://doi.org/10.1007/978-981-92-0176-1_8

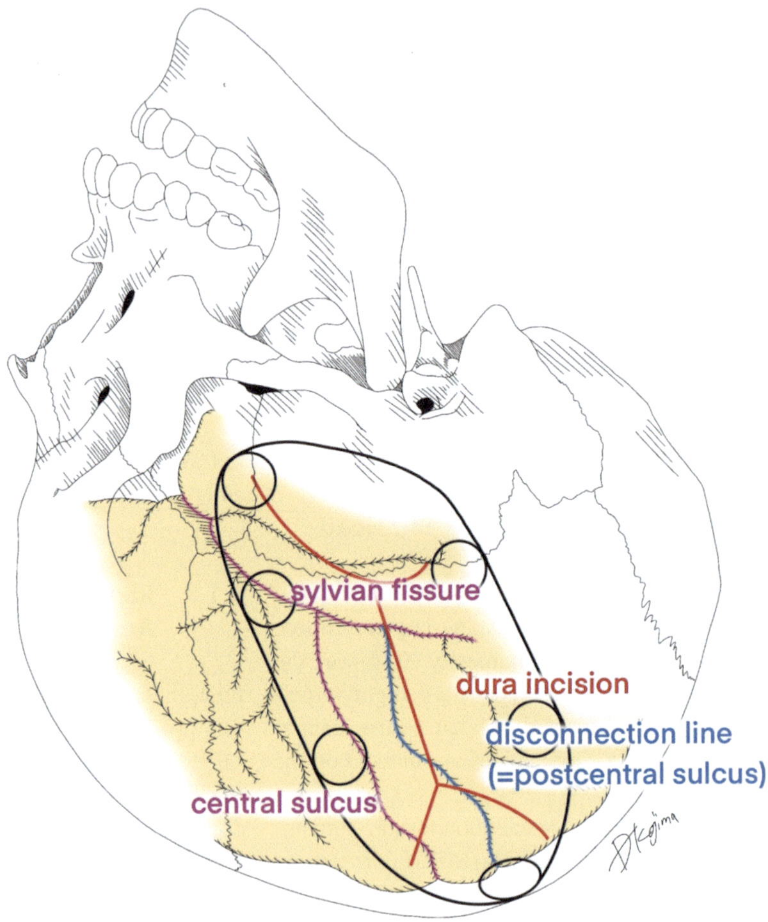

Fig. 8.1 Craniotomy for posterior quadrant disconnection. Parieto-temporal craniotomy is performed to provide wide exposure from the proximal to distal Sylvian fissure, as well as along the entire length of the postcentral sulcus

point, disconnection is advanced along the boundary between the hippocampal head and the uncus, while the ambient cistern and its contents—including the anterior choroidal artery and the oculomotor nerve—are visualized through the pia. By connecting this medial disconnection to the anterior temporal disconnection, the anterior temporal disconnection is completed. The incision through the temporal stem is then extended posteriorly along the posterior cerebral artery (PCA), opening the atrium from the temporal horn.

8.1.5 Parietal and Occipital Disconnection

Parietal cortical disconnection is performed along the postcentral sulcus. Medially, the disconnection is advanced toward the falx as a deep landmark until the interhemispheric fissure is reached. After entering the body of the lateral ventricle, the medial wall of the ventricular body is disconnected, and the posterior cingulate gyrus is transected, proceeding to the surface of the corpus callosum with the inferior edge of the falx as a landmark. This parietal disconnection is then connected to the previously created temporal disconnection (Fig. 8.2).

From the isthmus to the splenium of the corpus callosum, disconnection of the medial ventricular wall is advanced along its curvature. Using the inferior edge of the falx as a landmark, the disconnection is extended to a level beyond the falcotentorial junction. Within the atrium, the choroid plexus is elevated to expose the fornix, which is subsequently transected. Disconnection is further extended through the posterior parahip-

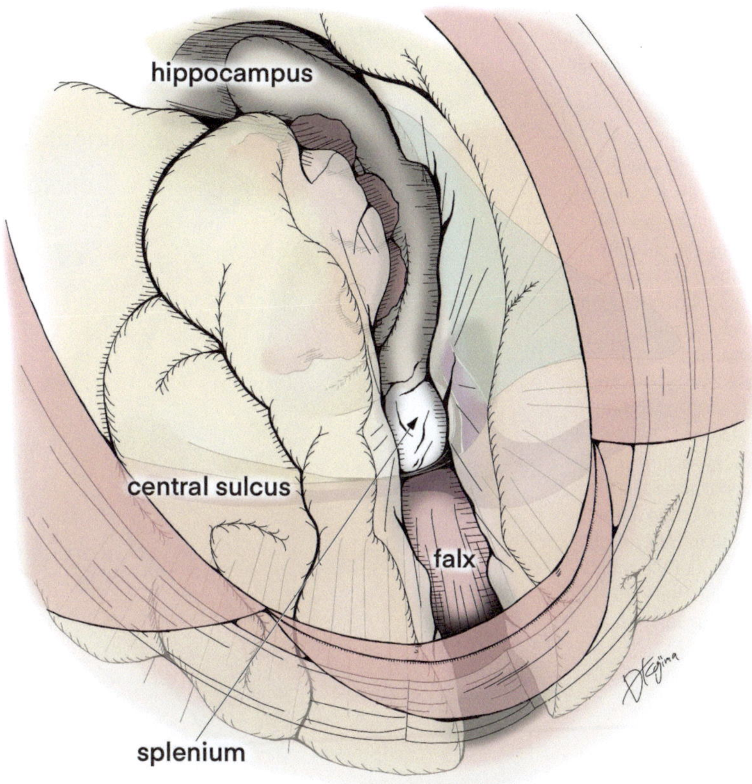

Fig. 8.2 Parietal and occipital disconnection. Parietal cortical disconnection is performed along the postcentral sulcus. Medially, the disconnection is advanced toward the falx as a deep landmark until the interhemispheric fissure is reached. After entering the body of the lateral ventricle, the medial wall of the ventricular body is aspirated, and the posterior cingulate gyrus is transected, proceeding to the splenium of the corpus callosum with the inferior edge of the falx as a landmark

pocampal gyrus and connected to the disconnection at the splenium (Fig. 8.3). With these steps, complete disconnection of the posterior quadrant—including the temporal, parietal, and occipital lobes—is achieved.

8.1.6 Closure

After confirming hemostasis, the dura is closed, the wound is closed in layers, and the procedure is completed.

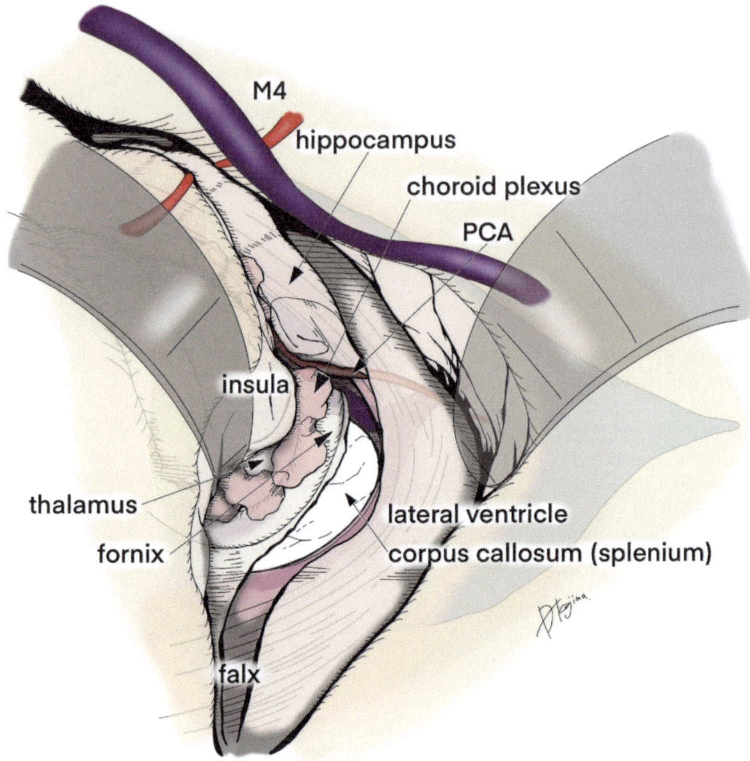

Fig. 8.3 Splenial and fornicial disconnection. Using the inferior edge of the falx as a landmark, the disconnection is extended to a level beyond the falcotentorial junction. Within the atrium, the choroid plexus is elevated to expose the fornix, which is subsequently transected. Disconnection is further extended through the posterior parahippocampal gyrus and connected to the disconnection at the splenium (*M4* fourth segment of the middle cerebral artery, *PCA* posterior cerebral artery)

8.2 Note

8.2.1 Surgical Outcomes and Procedural Variations

Posterior quadrant disconnection (PQD) has evolved to reduce surgical morbidity while maintaining the favorable seizure outcomes of posterior quadrantectomy in patients with drug-resistant epilepsy and extensive lesions involving the temporal, parietal, and occipital lobes [1, 2]. Compared with posterior quadrantectomy, PQD shares the general advantages of disconnection procedures, including shorter operative time and fewer surgical complications such as brain shift.

Seizure outcomes after PQD have been reported to vary widely, ranging from 25% to 90% across studies. In the largest PQD case series to date, Rizzi et al. reported outcomes in 42 patients, with seizure freedom (Engel class I) achieved in 34/42 patients (80.9%) at 6 months, 29/39 (74.5%) at 2 years, and 25/34 (73.5%) at 5 years postoperatively [3].

The surgical procedure of PQD is not standardized. The standard procedure, originally reported by Daniel et al. in 2007 and subsequently described in numerous reports, involves complete disconnection of the temporoparietooccipital region, with the postcentral sulcus serving as the anterosuperior limit of disconnection [1, 2]. However, it is preferable to tailor the disconnection according to the localization of the epileptogenic zone (EZ) and functional mapping, particularly in the language-dominant hemisphere. Temporal, parietal, and occipital disconnection procedures refer to performing part of the PQD disconnection steps described above. Overall, these reports suggest that the extent of disconnection in PQD should be individualized based on each patient's condition.

8.2.2 Complications

Relatively common complications after PQD include transient or permanent hemiparesis, as well as visual field deficits, including hemianopia and quadrantanopia [3–5]. In patients with preoperative visual field deficits, postoperative visual field findings often remain unchanged.

Most postoperative hemiparesis is transient and often improves over time. This deficit is thought to be primarily related to manipulation near the sensorimotor cortex and/or effects on the corticospinal tract within the subcortical white matter [5]. To avoid or mitigate such motor deficits, intraoperative functional monitoring using SEPs and MEPs is informative.

In a systematic review including 137 patients undergoing total or partial PQD, additional complications were reported, including weakness due to corticospinal tract injury, parietal lobe dysfunction, and aphasia associated with surgery in the dominant hemisphere. These functional deficits were often already present preoperatively. Considering the overall neurocognitive benefit associated with seizure freedom after PQD, the authors noted that "the benefits of seizure freedom on neurocognition may justify the potential risks." [6]

In particular, PQD involving the language-dominant hemisphere carries a higher risk of language dysfunction. Therefore, the preoperative identification of language lateralization and language-related regions using functional MRI (fMRI) and the Wada test is important. In addition, depending on the case, intraoperative monitoring of brain function with awake surgery and/or CCEPs can also be helpful.

8.2.3 Neurocognitive Outcomes

Because PQD involves extensive disconnection, including the parietal lobe, posterior temporal lobe, and occipital lobe, postoperative effects on higher-order brain functions related to the disconnected regions may occur. However, few studies have systematically evaluated long-term neurocognitive outcomes after PQD. Lippé et al. assessed neurocognitive function over 3–7 years after occipital and parietal disconnection in five pediatric patients with drug-resistant epilepsy due to focal cortical dysplasia. It reported improvement in visual perceptual cognition, although the degree of improvement was limited compared with the recovery of language function [7].

We identified the seizure onset zone (SOZ) using SEEG and performed nondominant PQD after the functional mapping of visuospatial cognition. Seizure outcome and neurocognitive function were evaluated over a 2-year postoperative period. The patient achieved Engel class I seizure outcome and showed an overall improvement on the Wechsler Adult Intelligence Scale-Fourth Edition (WAIS)-IV. In addition, no decline was observed in face recognition or topographic recognition [8].

At present, detailed reports evaluating long-term neurocognitive outcomes after PQD remain limited. However, existing evidence suggests that PQD can contribute to favorable seizure control and that with appropriate functional mapping, neurophysiological monitoring, and selection of the disconnection extent accordingly, the risk of permanent neurocognitive impairment may be relatively limited.

8.2.4 Subtotal Hemispherotomy

The combination of PQD and frontal lobe disconnection (FD), as described in Chap. 7, is referred to as subtotal hemispherotomy (SH) [9]. In SH, the precentral gyrus, postcentral gyrus, and insula remain connected (Fig. 8.4). This procedure is indicated for cases in which the epileptogenic zone is considered to spare the central region or when hemiparesis resulting from hemispheric disconnection is not tolerable. SH may be performed as an additional procedure after corpus callosotomy, FD, and PQD, but it can also be performed as a single-stage operation.

The surgical procedure for SH is as follows. The central sulcus and primary motor cortex are identified, and SEPs and MEPs are used to preserve motor function. When SH is performed as a

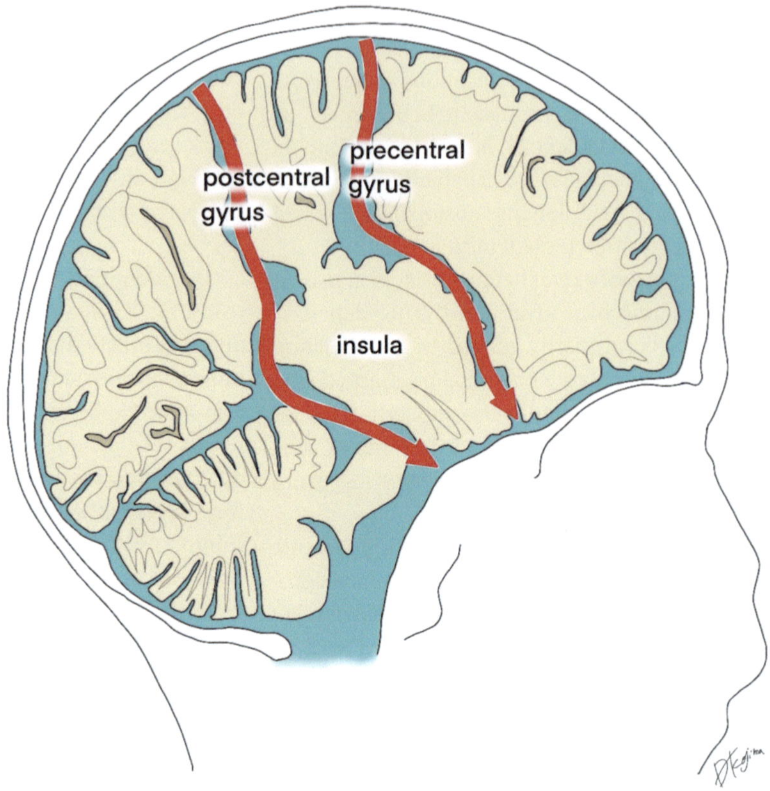

Fig. 8.4 Illustration of subtotal hemispherotomy. Subtotal hemispherotomy is a combination of posterior quadrant disconnection and frontal lobe disconnection. In SH, the precentral gyrus, postcentral gyrus, and insula remain connected

single-stage operation, the head is fixed in a horizontal position, and a frontotemporoparietal craniotomy is created to include the Sylvian fissure from proximal to distal, as well as the full length of the precentral and postcentral sulci. After craniotomy and dural opening, the procedures for FD and PQD are performed. For details regarding FD, please refer to the previous chapter.

In the case of additional central callosotomy after FD and PQD, access is obtained through the lateral ventricle. The precentral and postcentral gyri are gently retracted anteriorly and posteriorly, and an incision is made in the medial wall of the lateral ventricular body. Direct visualization of the pericallosal artery through the pia confirms the disconnection of the callosal fibers (Fig. 8.5).

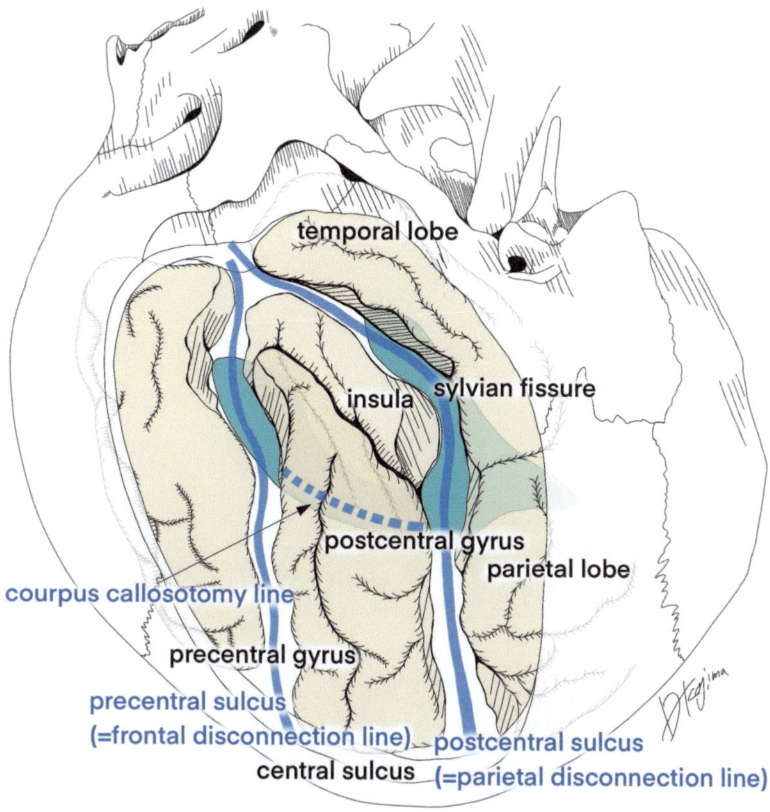

Fig. 8.5 Craniotomy and disconnection lines for subtotal hemispherotomy. A large frontotemporoparietal craniotomy is created to include the Sylvian fissure from proximal to distal, as well as the full length of the precentral and postcentral sulci. Frontal lobe disconnection and posterior quadrant disconnection are performed (solid blue lines). In the case of additional central callosotomy, the precentral and postcentral gyri are gently retracted anteriorly and posteriorly, and an incision is made in the medial wall of the lateral ventricular body (dotted blue line)

References

1. Umaba R, Uda T, Nakajo K, Kawashima T, Tanoue Y, Koh S, et al. Anatomic understanding of posterior quadrant disconnection from cadaveric brain, 3D reconstruction and simulation model, and intraoperative photographs. World Neurosurg. 2018;120:e792–801. https://doi.org/10.1016/j.wneu.2018.08.168.
2. Daniel RT, Meagher-Villemure K, Farmer JP, Andermann F, Villemure JG. Posterior quadrantic epilepsy surgery: technical variants, surgical anatomy, and case series. Epilepsia. 2007;48(8):1429–37. https://doi.org/10.1111/j.1528-1167.2007.01095.x.
3. Rizzi M, Revay M, d'Orio P, Scarpa P, Mariani V, Pelliccia V, et al. Tailored multilobar disconnective epilepsy surgery in the posterior quadrant. J Neurosurg. 2020;132(5):1345–57. https://doi.org/10.3171/2019.1.JNS183103.
4. Davis KL, Murro AM, Park YD, Lee GP, Cohen MJ, Smith JR. Posterior quadrant epilepsy surgery: predictors of outcome. Seizure. 2012;21(9):722–8. https://doi.org/10.1016/j.seizure.2012.07.017.
5. D'Agostino MD, Bastos A, Piras C, Bernasconi A, Grisar T, Gross Tsur V, et al. Posterior quadrantic dysplasia or hemi-hemimegalencephaly: a characteristic brain malformation. Neurology. 2004;62(12):2214–20. https://doi.org/10.1212/01.wnl.0000130459.91445.91.
6. Markosian C, Dodson V, Zhang HJ, Mahalingam RS, Geller EB, Tomycz LD. Total and partial posterior quadrant disconnection for medically refractory epilepsy: a systematic review. Seizure. 2021;91:66–71. https://doi.org/10.1016/j.seizure.2021.05.018.
7. Lippé S, Bulteau C, Dorfmuller G, Audren F, Delalande O, Jambaqué I. Cognitive outcome of parietooccipital resection in children with epilepsy. Epilepsia. 2010;51(10):2047–57. https://doi.org/10.1111/j.1528-1167.2010.02651.x.
8. Kawase N, Nakae S, Kumon M, Hayakawa M, Shinzato C, Sato Y, et al. Evaluation of higher cognitive functions following posterior quadrant disconnection in the non-dominant hemisphere: a case report. Front Hum Neurosci. 2025;19:1569673. https://doi.org/10.3389/fnhum.2025.1569673.
9. Uda T, Kunihiro N, Koh S, Nakanishi Y, Nakajo K, Tanoue Y, et al. Anatomic understanding of subtotal hemispherotomy using cadaveric brain, 3-dimensional simulation models, and intraoperative photographs. Oper Neurosurg (Hagerstown). 2020;18(6):E209–18. https://doi.org/10.1093/ons/opz354.

9 Hemispherotomy

9.1 Surgical Procedures for Parasagittal Vertical Hemispherotomy

9.1.1 Patient Positioning and Head Fixation

The patient is placed in the supine position, with the head fixed in the neutral midline position. As in total or anterior corpus callosotomy, the head is slightly flexed (Fig. 6.2).

9.1.2 Skin Incision and Craniotomy

The extent of the craniotomy is determined using the coronal suture as the landmark. A craniotomy approximately 5 cm in diameter, extending about 3 cm anteriorly and 2 cm posteriorly from the coronal suture, is planned. The skin incision is made either as a linear incision over the coronal suture or as a curvilinear incision (Fig. 6.3). The dura is opened with a curvilinear incision and reflected medially, with the medial side as the base (Fig. 6.4).

9.1.3 Medial and Anterior Disconnection

Approximately a 3-cm square portion of the superior frontal gyrus is resected to gain access to the deep region. The body of the lateral ventricle is reached from the depth of the resection cavity. In the ventricle, the foramen of Monro and the choroid plexus are identified.

The cingulate gyrus is resected to reach the interhemispheric fissure. After entering the interhemispheric fissure, the pericallosal artery is identified (Fig. 9.1).

The pericallosal artery is first followed anteriorly in a retrograde manner. From the ventricular side, the genu and rostrum of the corpus callosum are sectioned. The disconnection is extended to the subcallosal area and advanced inferiorly toward the anterior skull base (Fig. 9.2). The dissection is subsequently carried laterally along the frontal base. The posterior part of the olfactory nerve is identified through the arachnoid membrane, confirming adequate basal disconnection (Fig. 9.3). The anterior disconnection is considered complete when the dissection reaches the proximal part of the Sylvian fissure (Fig. 9.4).

Subsequently, the pericallosal artery is traced posteriorly from the resection cavity of the superior frontal gyrus. The corpus callosum is sequentially sectioned from the body to the isthmus and then to the splenium. When dividing the posterior corpus callosum, the inferior edge of the falx serves as an anatomical landmark.

T. Uda et al., *Atlas of Epilepsy Surgery*, https://doi.org/10.1007/978-981-92-0176-1_9

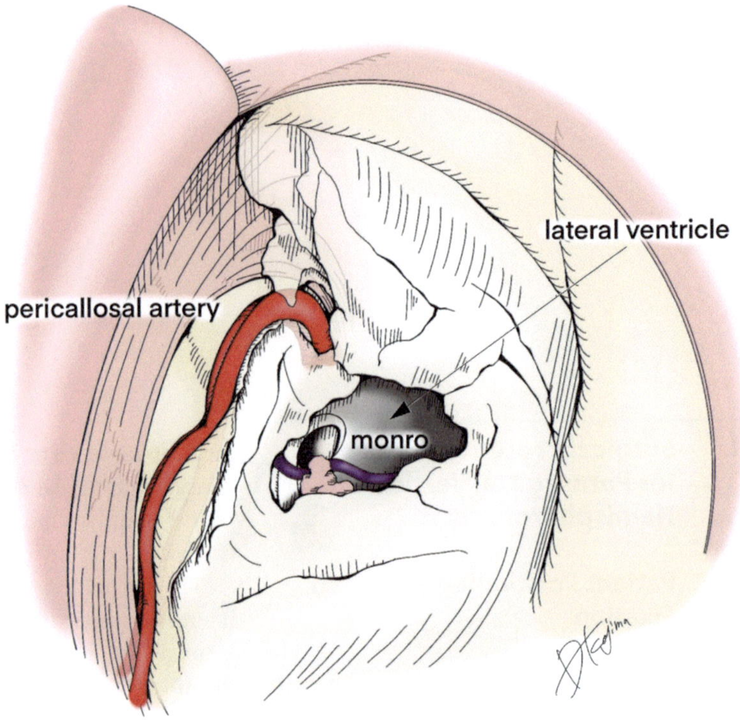

Fig. 9.1 Obtaining the surgical space for parasagittal vertical hemispherotomy. After a 3-cm square portion of the superior frontal gyrus is resected, the body of the lateral ventricle is opened through the resection cavity. In the ventricle, the foramen of Monro and the choroid plexus are identified. The cingulate gyrus is resected to reach the interhemispheric fissure, where the pericallosal artery is identified

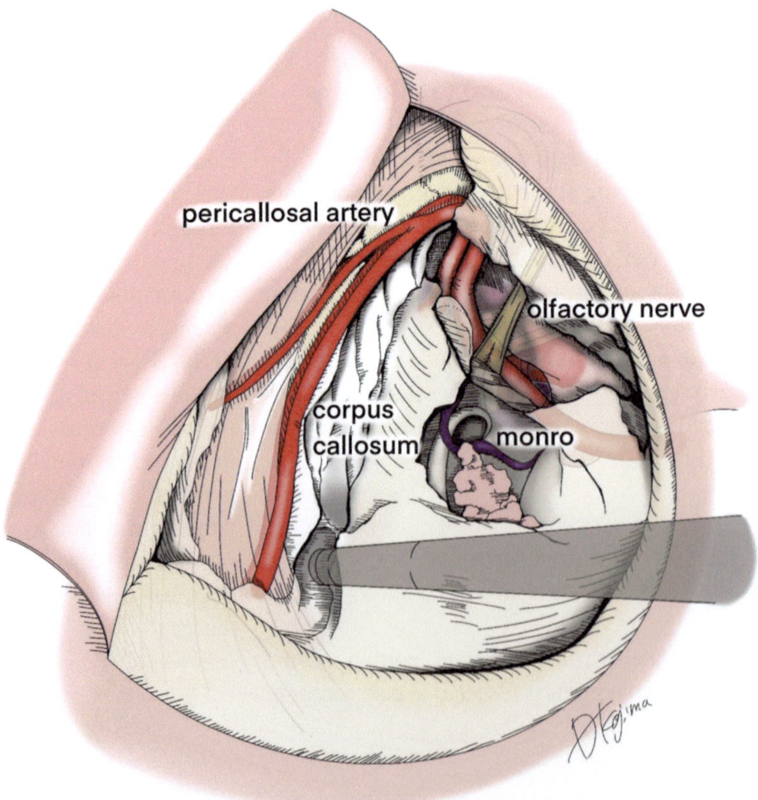

Fig. 9.2 Medial disconnection. Following the pericallosal artery, the genu and rostrum of the corpus callosum are sectioned in a transventricular fashion. After reaching the medial part of the anterior skull base, the dissection is subsequently carried laterally along the frontal base, where the posterior part of the olfactory nerve is identified

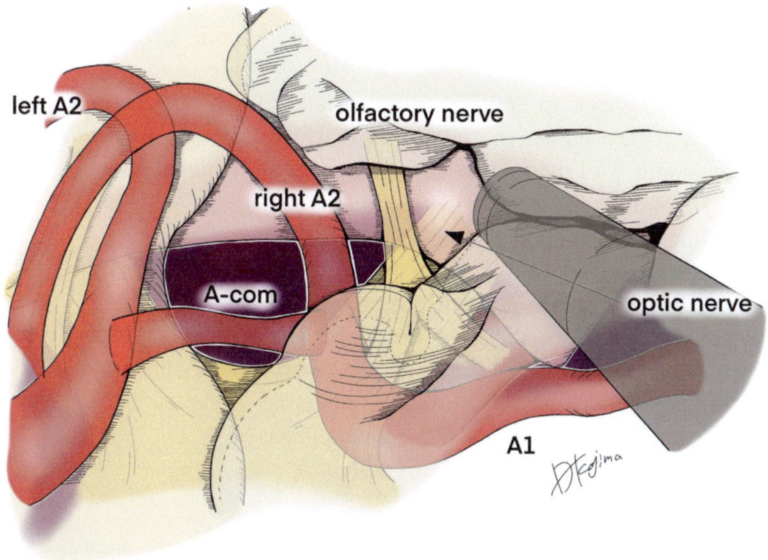

Fig. 9.3 Frontal base disconnection. The dissection is subsequently carried laterally along the frontal base. The posterior part of the olfactory nerve is identified through the arachnoid membrane (*A1* first segment of the anterior cerebral artery, *A2* second segment of the anterior cerebral artery, *A-com* anterior communicating artery)

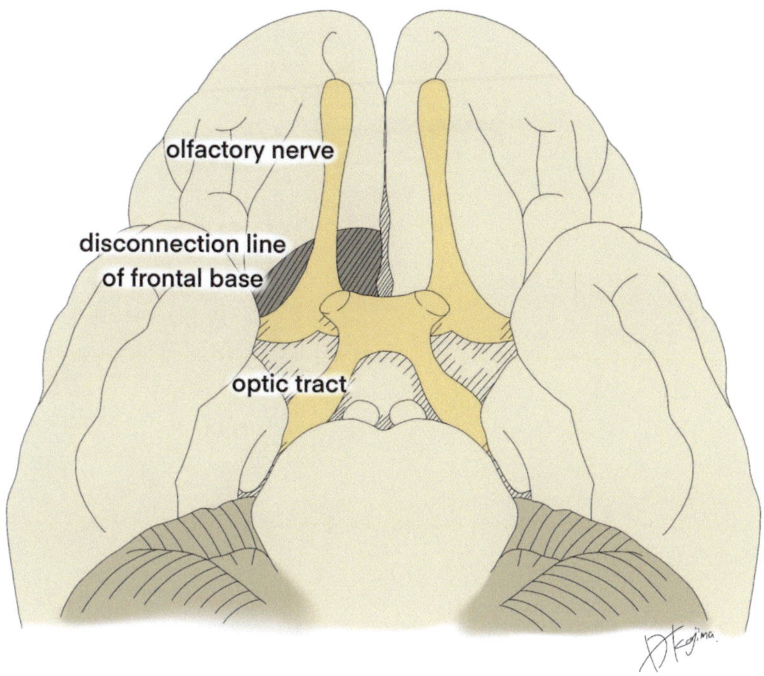

Fig. 9.4 Anteromedial disconnection. At the anteromedial part, the olfactory nerve is identified through the arachnoid membrane. The disconnection of the frontal base continued laterally until the proximal part of the Sylvian fissure was reached

9.1.4 Posterior and Lateral Disconnection

The choroid plexus in the trigone of the lateral ventricle is elevated to expose the fornix, which is continuous with the fimbria of the hippocampus. The fornix is sectioned, interrupting the ascending limbic projections from the hippocampus (Fig. 9.5).

Next, an incision is made within the lateral ventricle lateral to the thalamus (Fig. 9.6). The disconnection is then extended anteriorly toward the temporal horn, where the hippocampus is identified. The disconnection plane is extended between the caudate nucleus and the thalamus.

From the inferior choroidal point, mesial temporal disconnection continued anteriorly, extending beyond the proximal Sylvian fissure to the frontal base and connecting with the previously created dissection plane, completing the circumferential disconnection (Fig. 9.7).

9.1.5 Closure

After confirming hemostasis, the dura is closed with sutures, and the bone flap is replaced to complete the procedure.

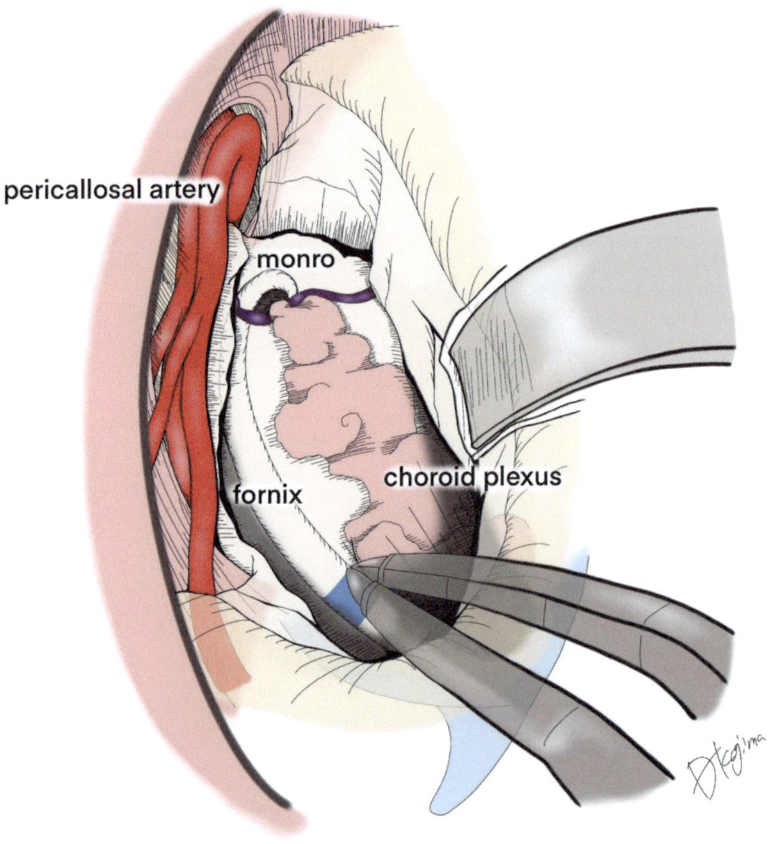

Fig. 9.5 Disconnection of the fornix. The choroid plexus in the trigone of the lateral ventricle is elevated to expose the fornix. The fornix is sectioned, interrupting the ascending limbic projections from the hippocampus

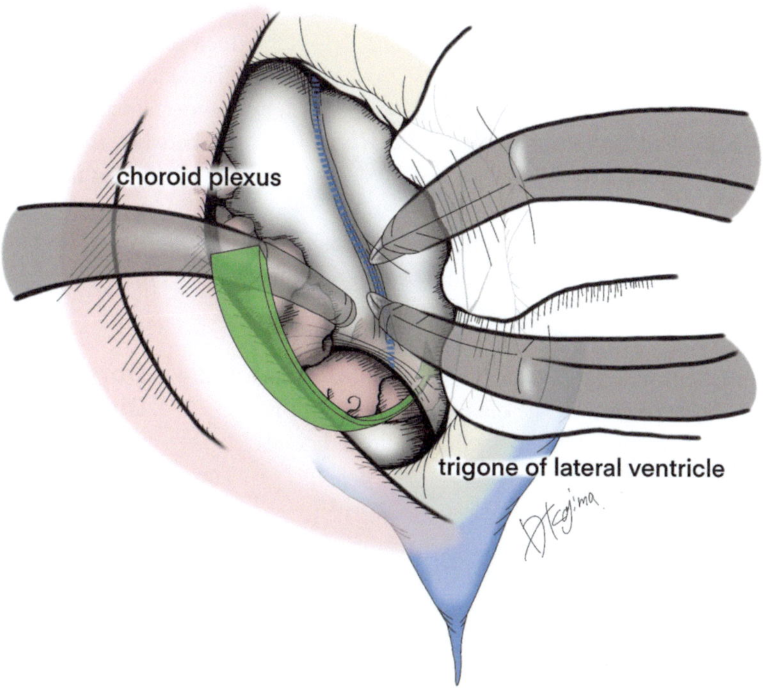

Fig. 9.6 Disconnection of the thalamus. An incision is made in the trigone of the lateral ventricle, lateral to the thalamus. The disconnection is then extended anteriorly toward the temporal horn

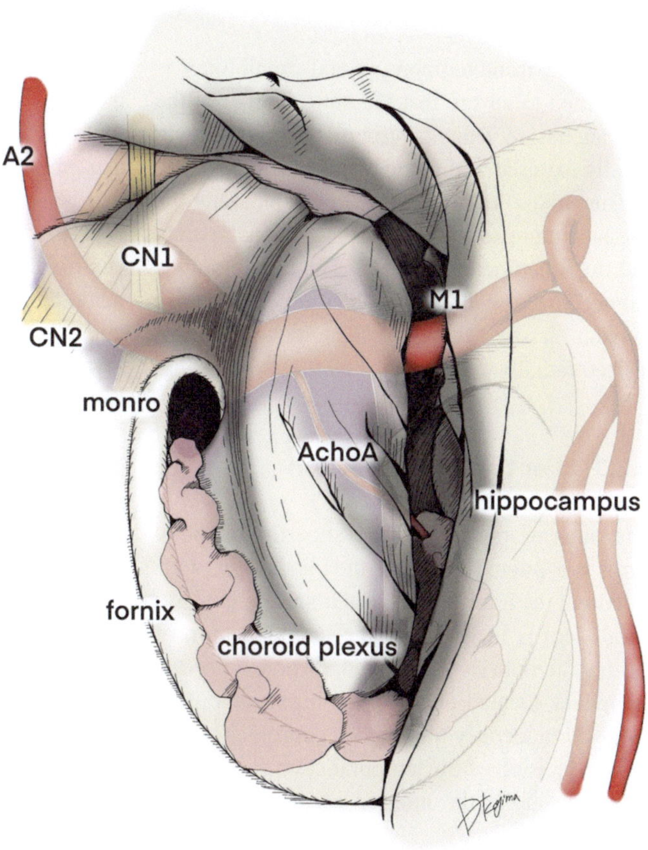

Fig. 9.7 Disconnection of the mesial temporal lobe toward the frontal base. The disconnection is then extended anteriorly toward the temporal horn, where the hippocampus is identified. From the inferior choroidal point, mesial temporal disconnection continued anteriorly and extended beyond the proximal Sylvian fissure to the frontal base and connected with the previously created dissection plane, completing the circumferential disconnection

9.2 Surgical Procedures of the Lateral Hemispherotomy

9.2.1 Patient Positioning and Head Fixation

The patient is placed in a supine-lateral position. A shoulder roll is inserted to allow slight rotation of the trunk, and the head is rotated approximately 80° toward the contralateral side. In lateral hemispherotomy, a lateral approach is employed; therefore, the operative hemisphere is positioned uppermost to facilitate direct lateral access.

9.2.2 Skin Incision and Craniotomy

A question-mark-shaped skin incision is made, extending from the anterior to the auricle, curving posteriorly around it, and then proceeding superiorly toward the frontal region. The scalp flap and temporalis muscle are reflected anteriorly as a single layer.

A wide frontotemporoparietal craniotomy is performed to ensure that the entire length of the Sylvian fissure, from its proximal to distal extent, is fully exposed within the operative field.

Burr holes are placed at the pterion, along the coronal suture, at the root of the zygomatic arch, and over the linea temporalis. The craniotomy is tailored to allow direct visualization of the entire Sylvian fissure (Fig. 9.8). The dura mater is incised radially.

9.2.3 Ventral Disconnection

The anterior portion of the superior temporal gyrus is aspirated to expose the insula (Fig. 9.9).

Along the inferior periinsular sulcus, the temporal stem is incised, and the temporal horn of the lateral ventricle is entered. A portion of the amygdala, located opposite the hippocampus, is removed to identify the inferior choroidal point.

The incision of the temporal stem is extended anteriorly along the inferior periinsular sulcus, sectioning the uncinate fasciculus. The uncus is then reached and aspirated. From this point, the disconnection is directed posteriorly toward the inferior choroidal point, interrupting the continuity between the medial temporal structures and the basal ganglia.

The incision of the temporal stem is extended posteriorly toward the trigone of the lateral ventricle (Fig. 9.10).

9.2.4 Dorsal and Interhemispheric Disconnection

The dissection proceeds around the distal Sylvian fissure toward the parietal lobe. The parietal and frontal opercula are resected along the superior periinsular sulcus. The lateral ventricle is subsequently entered, and the operative field is shifted to the body of the lateral ventricle (Fig. 9.11).

Further resection of the inferior frontal gyrus (frontal operculum) is carried anteriorly, reaching the frontal base at the sphenoid ridge. Advancing medially along the sphenoid ridge, the olfactory nerve is identified. Posterior to the olfactory nerve, the optic nerve, and the A1 segment of the anterior cerebral artery may be identified through the arachnoid membrane. The dissection is then continued medially to reach the interhemispheric fissure (Fig. 9.12).

The procedure is then transitioned to the transventricular callosal disconnection. An incision is made along the medial wall of the body of the lateral ventricle to reach the interhemispheric surface. The inferior edge of the falx and the pericallosal artery are identified.

The posterior corpus callosum is first completely sectioned up to the splenium. In proximity to the falcotentorial junction, the dissection is then extended inferiorly to section the fornix and the parahippocampal gyrus toward the tentorial incisura. Through this process, the posterior callosal fibers and the ascending limbic fibers from the hippocampus are interrupted.

The disconnection is then advanced anteriorly along the inferior edge of the falx, following the pericallosal artery in a retrograde manner. The genu of the corpus callosum is sectioned by curv-

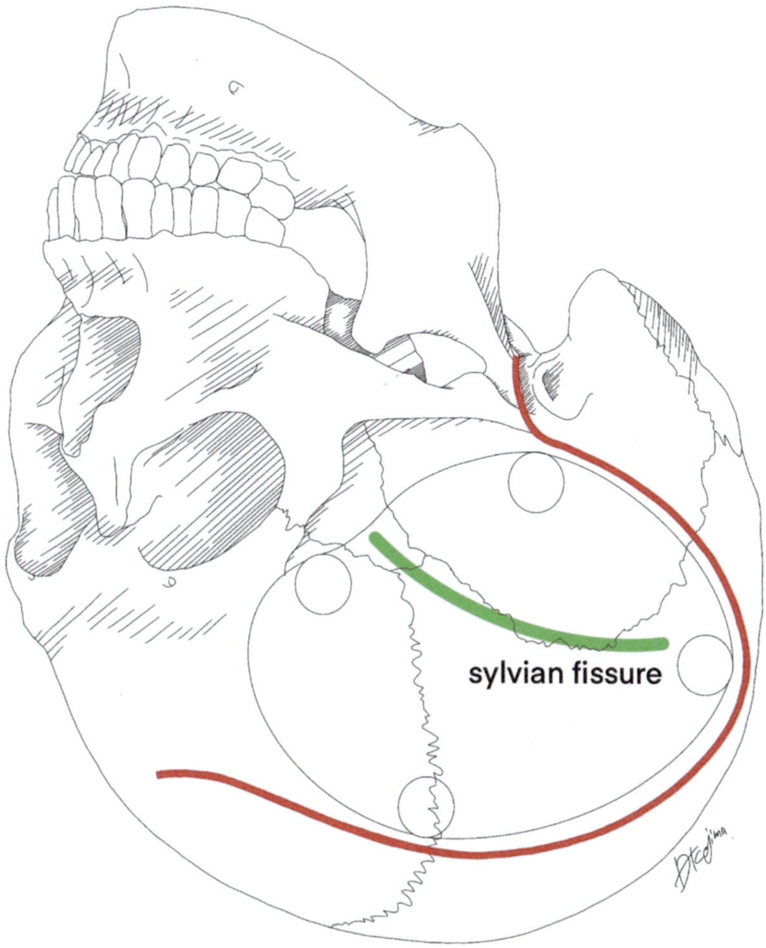

Fig. 9.8 Skin Incision and craniotomy for lateral hemispherotomy. A question mark-shaped skin incision is made from the anterior to the auricle, curving posteriorly around it, and then proceeding superiorly toward the frontal region. A wide frontotemporoparietal craniotomy is performed to expose the entire length of the Sylvian fissure. Burr holes are placed at the pterion, along the coronal suture, at the root of the zygomatic arch, and over the linea temporalis

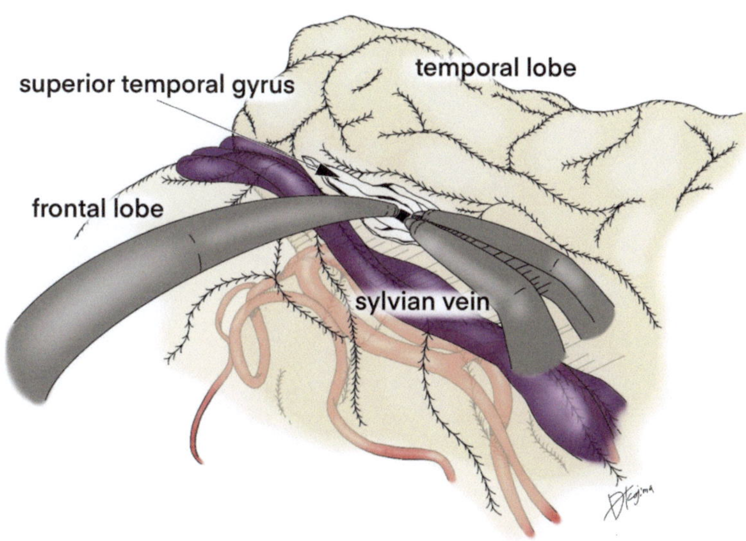

Fig. 9.9 Cortical incision of the temporal lobe. After the dural incision, the anterior portion of the superior temporal gyrus is aspirated to expose the insula

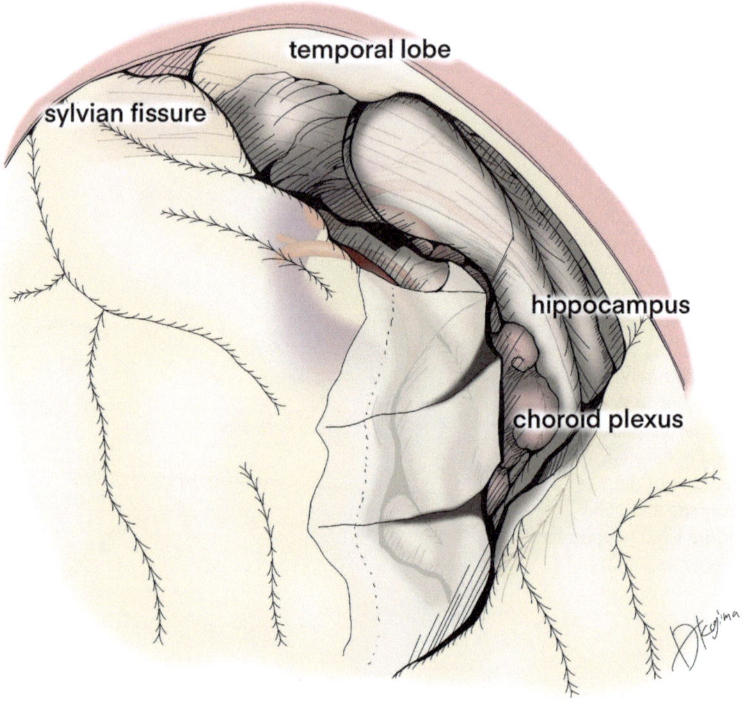

Fig. 9.10 Disconnection of projection fibers through the temporal operculum. After entering the inferior horn of the lateral ventricle, a part of the amygdala is removed to identify the inferior choroidal point. The incision of the temporal stem is extended anteriorly, and the uncus is aspirated. Then the disconnection is directed posteriorly toward the inferior choroidal point, interrupting the continuity between the medial temporal structures and the basal ganglia. The incision of the temporal stem is extended posteriorly toward the trigone of the lateral ventricle

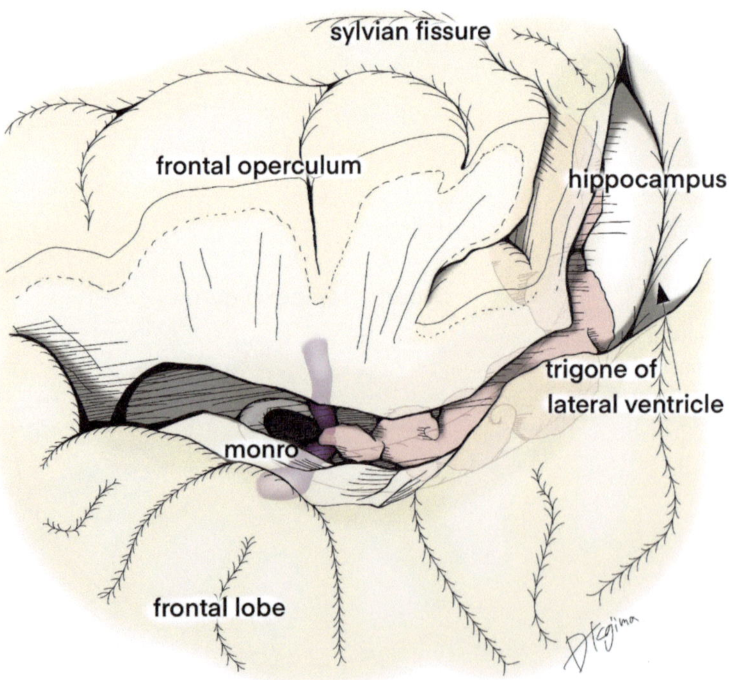

Fig. 9.11 Disconnection of projection fibers through the parietal and frontal operculum. The disconnection proceeds around the distal Sylvian fissure toward the parietal operculum, then the frontal operculum along the superior periinsular sulcus

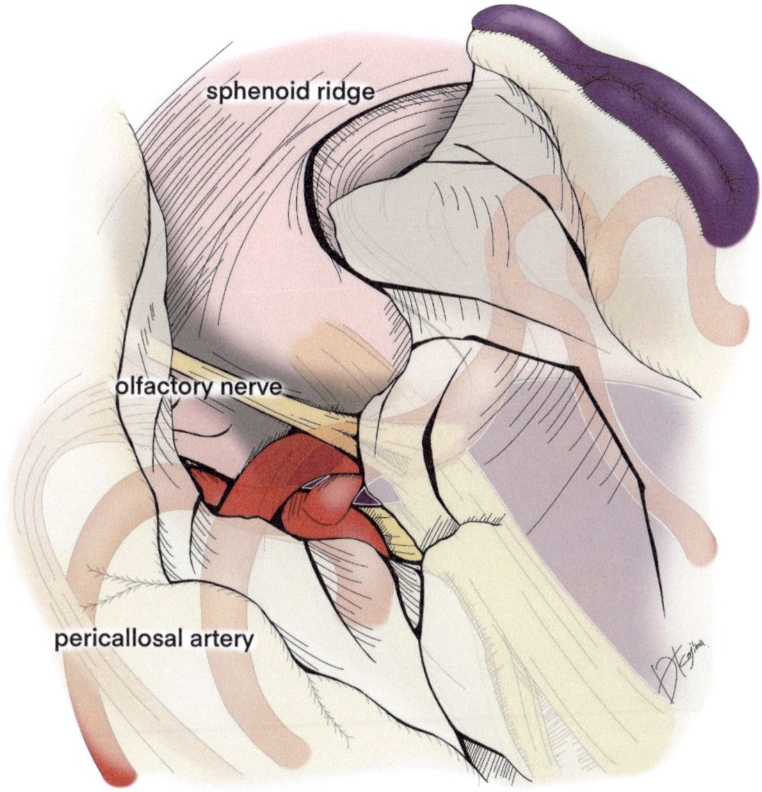

Fig. 9.12 Disconnection of the frontal base. The frontal operculum is resected anteriorly until the frontal base at the sphenoid ridge is reached. Advancing medially along the sphenoid ridge, the olfactory nerve is identified. Posterior to the olfactory nerve, the optic nerve, and the A1 segment of the anterior cerebral artery may be identified through the arachnoid membrane. The dissection is then continued medially to reach the interhemispheric fissure

ing around it along the course of the anterior cerebral artery, reaching the anterior communicating artery and the A1 segment. This medial disconnection is connected with the previously established lateral disconnection route.

9.2.5 Confluence of the Ventral and Dorsal Disconnection Planes

Finally, the insula is aspirated from both the body and temporal horn aspects of the lateral ventricle, creating continuity between the dorsal and ventral disconnection planes.

9.2.6 Closure

After confirming hemostasis, the dura is closed with sutures, and the bone flap is replaced to complete the procedure.

9.3 Notes

9.3.1 Comparison of Surgical Approaches

Illustrations of vertical hemispherotomy and lateral hemispherotomy were demonstrated (Fig. 9.13 and Fig. 9.14). The differences between these procedures lie in the size of the craniotomy and the disconnective approach to the projection fibers and the insula. In vertical hemispherotomy, projection fibers are disconnected from the lateral ventricle. By proceeding with this procedure, insular disconnection is simultaneously completed [1]. On the other hand, in lateral hemispherotomy, projection fibers are disconnected from the periinsular surface. Insular disconnection is subsequently performed separately.

Hemispherotomy has been reported to provide superior seizure outcomes and a lower incidence of postoperative complications compared with anatomical hemispherectomy.

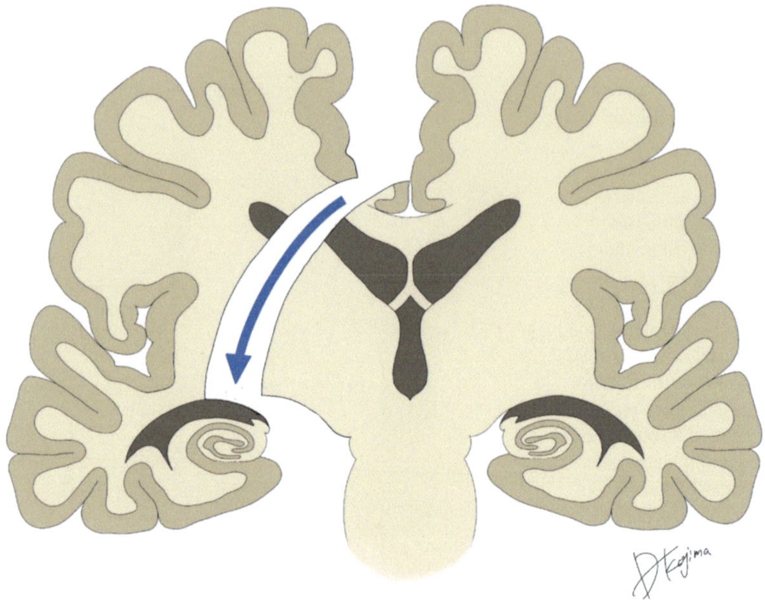

Fig. 9.13 Schematic drawing of vertical hemispherotomy. Projection fibers are disconnected from the lateral ventricle. By proceeding with this procedure, insular disconnection is simultaneously completed

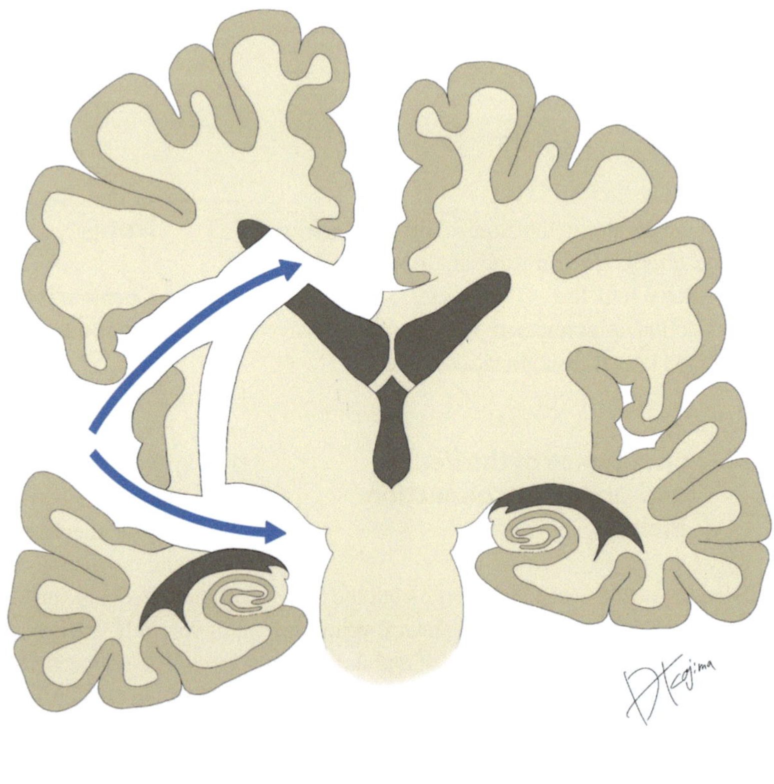

Fig. 9.14 Schematic drawing of lateral hemispherotomy. Projection fibers are disconnected from the periinsular surface. Insular disconnection is subsequently performed separately

In a meta-analysis by Cossu et al., the rate of Engel class I outcomes was approximately 80% for both the lateral and vertical approaches, with no statistically significant differences between the two groups. No significant differences were observed in the incidence of major complications or shunt-dependent hydrocephalus [2].

In contrast, an individual patient data meta-analysis by Chen et al. focusing on pediatric drug-resistant epilepsy (DRE) demonstrated that the vertical approach was associated with a higher rate of seizure freedom and greater long-term durability compared with the lateral approach, with comparable safety profiles [3]. The vertical approach was associated with fewer seizure recurrences requiring reoperation, a higher rate of seizure freedom at final follow-up, and a significantly longer time to seizure recurrence. Post hoc analyses from the same group further supported the superiority of the vertical approach in achieving sustained long-term seizure freedom. Moreover, reported 5- and 10-year seizure freedom rates were approximately 70% and 50–60%, respectively, for the lateral approach, compared with approximately 80–90% and 85%, respectively, for the vertical approach [4, 5]. Notably, Chen et al. minimized bias inherent to retrospective studies through methods such as propensity score matching, providing important evidence that the observed differences are attributable to procedural characteristics rather than baseline patient factors.

Comparative studies of anatomical hemispherectomy and functional hemispherotomy have shown no significant difference in seizure outcomes; however, hydrocephalus occurs more frequently following anatomical hemispherectomy [6]. In cases of hemimegalencephaly, marked anatomical distortion may hinder complete disconnection using functional techniques, and a higher reoperation rate has been reported after functional hemispherotomy in this population [7].

9.3.2 Complications of Hemispherotomy

In a large review of 1161 cases of hemispherectomy and hemispherotomy by Griessenauer et al., the most common complication was hydrocephalus requiring cerebrospinal fluid shunt placement, occurring in 14% of cases overall [6]. The incidence was highest after anatomical hemispherectomy and significantly lower following functional hemispherectomy and hemispherotomy. Reported shunt rates in the literature range from 8% to 23% and appear to correlate with the volume of resected brain tissue.

As with other intracranial procedures, postoperative complications may include infection, aseptic meningitis, and transient neurological deficits. Contralateral cerebral infarction and perioperative mortality are rare in contemporary series, with mortality rates estimated at less than 1% in experienced centers [8].

A systematic review and meta-analysis by Karagianni et al. also demonstrated that less invasive hemispherotomy techniques are associated with lower complication rates compared with conventional anatomical hemispherectomy [9]. Anatomical hemispherectomy, which involves extensive multilobar resection and substantial intraoperative blood loss, carries a higher risk of mortality and major complications. In contrast, reported mortality rates associated.

with hemispherotomy range from 0% to 1.0% [10].

Neurological complications are generally infrequent regardless of the surgical technique. Worsening of preexisting hemiparesis or new-onset hemiparesis has been reported in 8–21% of cases. In adult patients, postoperative language decline occurs in approximately 10% of cases, particularly after dominant-hemisphere surgery [11].

9.3.3 Endoscopic Hemispherotomy

Hemispherotomy for drug-resistant epilepsy has traditionally required extensive craniotomy and has been considered highly invasive. In recent years, endoscope-assisted functional hemispherotomy has been introduced to reduce surgical morbidity.

Wagner et al., in a cadaveric study, demonstrated that all steps of hemispheric disconnection can be accomplished using standard microsurgical instruments and a rigid endoscope. Clinical application in two pediatric patients confirmed complete hemispheric disconnection on postoperative tractography [12]. The authors concluded that endoscope-assisted functional hemispherotomy is technically feasible and may represent a safe, minimally invasive alternative in appropriately selected patients. Compared with conventional microscopic techniques, this approach may reduce surgical trauma and facilitate faster postoperative recovery.

In a prospective observational study, Chandra et al. compared endoscope-assisted hemispherotomy (EH) with conventional open hemispherotomy (OH). Among 59 patients, EH demonstrated seizure and quality-of-life outcomes comparable to OH [13]. Additionally, EH was associated with reduced intraoperative blood loss and shorter hospital stay.

We have also reported endoscopic parasagittal vertical hemispherotomy for pediatric drug-resistant epilepsy [14]. This minimally invasive approach, performed through a small craniotomy, enables precise and safe hemispheric disconnection. No postoperative complications were observed, and no seizure recurrence occurred during long-term follow-up. Collectively, these findings suggest that endoscopic hemispherotomy is a promising technique in terms of both safety and feasibility.

References

1. Uda T, Tamrakar S, Tsuyuguchi N, Kawashima T, Goto H, Nakajo K, et al. Anatomic understanding of vertical hemispherotomy with cadaveric brains and intraoperative photographs. Oper Neurosurg. 2016;12(4):374–82. https://doi.org/10.1227/NEU.0000000000001272.
2. Cossu M, Nichelatti M, De Benedictis A, Rizzi M. Commission for Epilepsy Surgery of the Italian league against epilepsy (LICE). Lateral versus vertical hemispheric disconnection for epilepsy: a systematic review and meta-analysis. J Neurosurg. 2022;136(6):1627–37. https://doi.org/10.3171/2021.5.JNS21949.
3. Chen JS, Harris WB, Wu KJ, Phillips HW, Tseng CH, Weil AG, et al. Comparison of hemispheric surgery techniques for pediatric drug-resistant epilepsy: an individual patient data meta-analysis. Neurology. 2023;101(4):e410–24. https://doi.org/10.1212/WNL.0000000000207425.
4. Fallah A, Lewis E, Ibrahim GM, Kola O, Tseng CH, Harris WB, et al. Comparison of the real-world effectiveness of vertical versus lateral functional hemispherotomy techniques for pediatric drug-resistant epilepsy: a post hoc analysis of the HOPS study. Epilepsia. 2021;62(11):2707–18. https://doi.org/10.1111/epi.17021.
5. Bourdillon P, Bulteau C, Dorfmuller G, Ferrand-Sorbets S. Vertical hemispherotomy for drug-resistant epilepsy: toward confirmation of the HOPS study. Epilepsia. 2021;62(12):3150–1. https://doi.org/10.1111/epi.17111.
6. Griessenauer CJ, Salam S, Hendrix P, Patel DM, Tubbs RS, Blount JP, et al. Hemispherectomy for treatment of refractory epilepsy in the pediatric age group: a systematic review. J Neurosurg Pediatr. 2015;15(1):34–44. https://doi.org/10.3171/2014.10.PEDS14155.
7. Di Rocco C, Battaglia D, Pietrini D, Piastra M, Massimi L. Hemimegalencephaly: clinical implications and surgical treatment. Childs Nerv Syst. 2006;22(8):852–66. https://doi.org/10.1007/s00381-006-0149-9.
8. Lew SM. Hemispherectomy in the treatment of seizures: a review. Transl Pediatr. 2014;3(3):208–17. https://doi.org/10.3978/j.issn.2224-4336.2014.04.01.
9. Karagianni MD, Brotis AG, Tasiou A, Delev D, von Lehe M, Schijns OEMG, et al. Hemispherotomy revised: a complication overview and a systematic

review meta-analysis. Brain Spine. 2023;3:101766. https://doi.org/10.1016/j.bas.2023.101766.
10. Schusse CM, Smith K, Drees C. Outcomes after hemispherectomy in adult patients with intractable epilepsy: institutional experience and systematic review of the literature. J Neurosurg. 2018;128(3):853–61. https://doi.org/10.3171/2016.9.JNS151778.
11. Schramm J, Delev D, Wagner J, Elger CE, von Lehe M. Seizure outcome, functional outcome, and quality of life after hemispherectomy in adults. Acta Neurochir. 2012;154(9):1603–12. https://doi.org/10.1007/s00701-012-1408-z.
12. Wagner K, Vaz-Guimaraes F, Camstra K, Lam S. Endoscope-assisted hemispherotomy: translation of technique from cadaveric anatomical feasibility study to clinical implementation. J Neurosurg Pediatr. 2019;23(2):178–86. https://doi.org/10.3171/2018.8.PEDS18349.
13. Chandra PS, Subianto H, Bajaj J, Girishan S, Doddamani R, Ramanujam B, et al. Endoscope-assisted (with robotic guidance and using a hybrid technique) interhemispheric transcallosal hemispherotomy: a comparative study with open hemispherotomy to evaluate efficacy, complications, and outcome. J Neurosurg Pediatr. 2019;23(2):187–97. https://doi.org/10.3171/2018.8.PEDS18131.
14. Koh S, Uda T, Kunihiro N, Umaba R, Ishimoto K, Kawashima T, et al. Endoscopic parasagittal vertical hemispherotomy: a 2-dimensional operative video. Oper Neurosurg (Hagerstown). 2025;29(4):598–9. https://doi.org/10.1227/ons.0000000000001466.

10 Intracranial Electrode (SEEG)

10.1 Surgical Procedures

10.1.1 Planning

Stereoelectroencephalography (SEEG) can be performed by using either a stereotactic frame-based technique or a robot-guided approach. At our institution, SEEG is performed using the Autoguide system (Medtronic, USA). For planning, three-dimensional images derived from contrast-enhanced computed tomography (CT), fluid-attenuated inversion recovery (FLAIR) magnetic resonance imaging (MRI), and contrast-enhanced MRI are imported into the StealthStation S8 navigation system.

The entry point is considered to be the most critical factor in SEEG planning. This is because hemorrhagic complications in SEEG are widely considered to originate at the cortical surface, making careful risk management at the entry point particularly important. Therefore, during planning, we confirm that no vascular structures are present within a 2.5 mm radius of the intended entry point. In addition, to minimize the potential reduction in electrode placement accuracy due to brain shift associated with cerebrospinal fluid (CSF) leakage, care is taken to ensure that the electrode trajectory does not contact the ventricular system.

Based on the presumed epileptic network and the findings of noninvasive investigations, the entry points and targets are determined to adequately sample the gyri suspected of involvement. These trajectories are then registered in the navigation system.

10.1.2 Patient Positioning and Fixation

The patient is placed in the supine position. To minimize positional discrepancies during electrode insertion, the head position is set as close as possible to that used during preoperative imaging. Specifically, the head is positioned with the vertex as flat as possible, and head rotation is avoided during fixation.

For head fixation, a carbon frame is used, as it minimizes metallic artifacts and allows clear fluoroscopic visualization of depth electrode placement.

10.1.3 Burr Hole Creation

Based on the registered planning data, the trajectory is determined using the Autoguide system. To prevent deviation of the insertion trajectory caused by scalp bulging, local anesthesia is not administered at the burr hole site. No skin incision is made; instead, a percutaneous burr hole is created using a drill.

Before burr hole creation, the distance from the scalp to the cortical surface at the entry point is measured, and the drilling depth is set accordingly. After burr hole creation, dural penetration

T. Uda et al., *Atlas of Epilepsy Surgery*, https://doi.org/10.1007/978-981-92-0176-1_10

is confirmed using the navigation system, and the cortical surface is coagulated.

10.1.4 Insertion of Depth Electrodes

First, a biopsy needle is inserted along the planned trajectory, and its tip position is marked on the fluoroscopic image. Subsequently, the depth electrode is inserted along the same trajectory under fluoroscopic guidance. After insertion, the electroencephalographic signals from each electrode are confirmed.

The depth electrodes are secured to the scalp using sutures, and fluoroscopy is repeated to confirm that the electrodes have neither been advanced excessively nor inadvertently withdrawn. Anchor bolts represent one of the standard fixation methods; however, their use depends on the facilities and resources available at each institution.

The procedures described in Sections 14.1.3 and 14.1.4 are repeated for the planned number of electrodes, after which the operation is completed.

10.1.5 Postoperative CT

A CT scan is obtained immediately after surgery to evaluate for hemorrhagic complications. In addition, fusion of postoperative CT and preoperative planning images allows confirmation that the electrodes follow the planned trajectories and reach the intended targets.

10.2 Note

10.2.1 Planning and Preparation of the Electrodes Table

During the planning process, an electrode table is prepared, and the procedure is performed according to this table (see Table 10.1). The electrode table includes, for each depth electrode, the contact numbers, the side of implantation, the entry gyrus, and the target gyrus.

In addition, the order of electrode insertion is predetermined during the planning stage. Because an increasing number of insertions may reduce electrode placement accuracy, electrodes near vascular structures are inserted first. In contrast, electrodes targeting structures adjacent to the ventricular system, such as the amygdala or parahippocampal gyrus, are inserted later in the procedure. This strategy is intended to minimize the impact on subsequent electrode placement accuracy should cerebrospinal fluid leakage occur due to ventricular contact.

10.2.2 Complications

Hemorrhagic complications represent the most common adverse events associated with SEEG. In a series of 549 SEEG procedures reported by McGovern et al., hemorrhagic complications were identified in 105 patients (19.1%). Of these, 93 cases (16.9%) were asymptomatic. In contrast, 12 patients (2.2%) experienced symptomatic hemorrhage, corresponding to an incidence of approximately 0.2% per electrode inserted [1]. In addition, several meta-analyses have reported a pooled prevalence of hemorrhagic complications of approximately 1–2% [2, 3].

Although the reported frequency of hemorrhagic complications in SEEG varies among studies, this variability is largely attributable to differences in outcome definitions. Most meta-analyses focus on clinically significant hemorrhage, and whether radiological hemorrhage is included or excluded can result in substantial differences in the reported incidence.

Hemorrhagic complications in SEEG may present in various forms, including intracerebral hemorrhage (ICH), subarachnoid hemorrhage (SAH), subdural hematoma (SDH), and epidural hematoma (EDH); among these, ICH has been reported to be the most frequent [2].

Infection is the second most common complication of SEEG. The pooled prevalence of infectious complications has been reported to be approximately 0.8%, with brain abscess as the most frequently observed manifestation. Other complications, including electrode-related mal-

Table 10.1 An example of electrode table used in our case

	Contacts	Side	Entry of implantation	Target of implantation
1	A1–A10	Right	Inferior temporal gyrus	Parahippocampal gyrus
2	A11–A20	Right	Inferior temporal gyrus	Parahippocampal gyrus
3	A21–A26	Right	Superior temporal gyrus	Temporal stem and insula
4	A27–A36	Left	Middle temporal gyrus	Parahippocampal gyrus (anterior)
5	A37–A46	Left	Middle temporal gyrus	Amygdala
6	A47–A56	Left	Middle temporal gyrus	Parahippocampal gyrus (posterior)
7	A57–A66	Right	Middle temporal gyrus	Amygdala

functions and permanent neurological deficits, have each been reported with pooled prevalence rates of less than 1% [2].

10.2.3 Comparison with Subdural Electrodes

In intracranial electroencephalogram (EEG) recording, SEEG and subdural electrodes (SDEs) differ in invasiveness, coverage, and functional mapping capabilities. Understanding the characteristics of each technique is essential for appropriate case selection.

From a surgical standpoint, SEEG involves the insertion of electrodes through burr holes approximately 2 mm in diameter and does not require craniotomy, resulting in a relatively low degree of invasiveness. In contrast, SDE placement requires craniotomy and is therefore associated with greater surgical invasiveness. Owing to these procedural differences, SEEG has been reported to be associated with shorter operative times compared with SDE implantation [3].

About recording coverage, SDEs allow extensive coverage of the cortical surface using grid electrodes, facilitating a two-dimensional assessment of cortical surface activity. In contrast, SEEG enables multidimensional recordings of the cortical surface, deep brain structures, noncontiguous cortical regions, and networks spanning both cerebral hemispheres. Consequently, SEEG is considered particularly useful for evaluating epilepsy involving deep-seated structures, suspected multifocal epilepsy, or epileptic networks that are difficult to assess using conventional SDE techniques [4].

In terms of functional mapping using electrical stimulation, SDEs are generally regarded as superior. Because SDEs provide broad and uniform coverage of the cortical surface, they are well suited for detailed stimulation mapping of eloquent cortical areas, such as language and motor cortices.

Both techniques offer high spatial resolution, although their strengths differ. SDEs excel in high-resolution, two-dimensional recordings along the cortical surface, whereas SEEG allows high-resolution, three-dimensional recordings that include deep brain structures. Furthermore, recent reviews have noted that the combined use of SEEG and SDEs in particular cases may enable more precise localization of the epileptogenic zone and more comprehensive functional mapping [4].

A systematic review and meta-analysis published in 2025, which included 15 studies, reported that postoperative seizure outcomes were 14% higher in the SEEG group than in the SDE group ($p = 0.02$) and that complication rates were significantly lower in the SEEG group ($p < 0.00001$) [3]. One proposed explanation is that SEEG electrode removal does not require repeat craniotomy, whereas SDE removal necessitates reoperation. As a result, when localization is insufficient, SDE implantation may introduce so-called surgical momentum, in which resection is pursued despite suboptimal localization. This phenomenon has been suggested to contribute to poorer seizure outcomes and higher complication rates.

In contrast, another meta-analysis published in 2025, which included 16 studies, reported no significant difference in seizure freedom rates between SEEG and SDE [5]. However, in that analysis, SEEG demonstrated statistically significant advantages over SDE in overall complication rates, severe hemorrhagic events, postoperative neurological deficits, and operative time. The authors concluded that "these findings indicate a possible preference for SEEG in clinical practice."

Taken together, these findings support SEEG as the predominant modality for intracranial EEG recording in contemporary epilepsy surgery.

10.2.4 Clinical Considerations and Limitations of SEEG

Intracranial electrode implantation is a diagnostic modality that enhances the spatial resolution of seizure onset zone (SOZ) localization and contributes to the identification of the epileptogenic zone, playing a critical role in the presurgical evaluation of focal epilepsy. Among these techniques, SEEG is a minimally invasive approach that does not require craniotomy and offers high diagnostic value. Nevertheless, the indication for intracranial electrode implantation, including SEEG, requires careful consideration on a case-by-case basis. Below, we outline clinical situations in which we do not actively recommend intracranial electrode implantation.

10.2.4.1 Patients with a Clear Structural Lesion and Infrequent Seizures

Gliomas and cerebral cavernous malformations (CCMs) are frequently associated with epilepsy. In patients with a well-defined structural lesion on MRI and infrequent seizures, intracranial electrode implantation is unlikely to capture seizures. In such cases, considering the balance between procedural invasiveness and potential diagnostic benefit, the additional value of SEEG is limited.

10.2.4.2 Patients with Frequent Seizures and Suspected High-Grade Glioma

An accurate determination of tumor grade based solely on imaging is often challenging. In treatment-naïve patients, molecular biological information is typically unavailable, thereby precluding complete exclusion of high-grade malignancy. Even when imaging findings suggest a low-grade glioma and seizures are frequent, these factors alone do not justify the selection of SEEG. Prioritizing SEEG in glioma-associated epilepsy may delay definitive surgical treatment and increase the risk of tumor progression. Therefore, indications for SEEG in patients undergoing initial treatment for glioma should be considered with caution.

Conversely, in tumors that demonstrate minimal growth over long-term radiological follow-up, the likelihood of rapid progression is low, and SEEG-based seizure focus localization may meaningfully contribute to treatment strategy.

10.2.4.3 Severe Coagulopathy

SEEG requires multiple burr holes and drilling procedures. As a result, there is a risk that persistent bleeding from multiple entry sites may propagate intracranially, as well as difficulty in achieving hemostasis at puncture sites. These factors may lead to serious intracranial hemorrhagic complications. Therefore, in patients with severe coagulopathy, the indication for SEEG should be assessed with extreme caution.

10.2.4.4 Cranial Bone Abnormalities, Suture Defects, or Localized Skull Defects

In patients with cranial discontinuity or morphological abnormalities—such as suture defects, cranial dysplasia, or localized skull defects—there is an increased risk of bone depression, separation, or fracture during head-pin fixation or burr hole creation. In such cases, secure head fixation is often difficult to achieve, and the performance of SEEG should be carefully reconsidered from a safety standpoint.

10.2.4.5 Generalized Epilepsy

SEEG is a technique predicated on focal seizure localization and is generally not indicated for patients with generalized epilepsy. However, particularly in pediatric cases, interictal scalp EEG findings may be nonspecific, and differentiation between focal and generalized epilepsies may not be straightforward [6]. Given the limitations of scalp EEG alone in determining seizure localization, SEEG may be considered only in carefully selected cases in which clinical semiology and imaging findings collectively suggest the possibility of focal epilepsy.

References

1. McGovern RA, Ruggieri P, Bulacio J, Najm I, Bingaman WE, Gonzalez-Martinez JA. Risk analysis of hemorrhage in stereo-electroencephalography procedures. Epilepsia. 2019;60(3):571–80. https://doi.org/10.1111/epi.14668.
2. Mullin JP, Shriver M, Alomar S, Najm I, Bulacio J, Chauvel P, et al. Is SEEG safe? A systematic review and meta-analysis of stereo-electroencephalography-related complications. Epilepsia. 2016;57(3):386–401. https://doi.org/10.1111/epi.13298.
3. Bandopadhay J, Rizzo N, Hanan C, Massoud AT, Pichardo-Rojas P, Esquenazi Y, et al. Comparative assessment of stereoelectroencephalography and subdural electrodes in invasive epilepsy monitoring: a systematic review and meta-analysis. J Neurosurg. 2025;143(5):1176–84. https://doi.org/10.3171/2025.3.JNS243188.
4. Wu S, Issa NP, Rose SL, Haider HA, Nordli DR Jr, Towle VL, et al. Depth versus surface: a critical review of subdural and depth electrodes in intracranial electroencephalographic studies. Epilepsia. 2024;65(7):1868–78. https://doi.org/10.1111/epi.18002.
5. Gomes FC, Larcipretti ALL, Udoma-Udofa OC, Rocha BAA, Mota MEB, Decina MM, et al. Stereoelectroencephalography versus subdural electrodes for invasive monitoring of drug-resistant epilepsy patients: a systematic review and meta-analysis. Seizure. 2025;129:33–41. https://doi.org/10.1016/j.seizure.2025.04.001.
6. Montenegro MA, Valente K. EEG in focal and generalized epilepsies: pearls and perils. Epilepsy Behav. 2024;156:109825. https://doi.org/10.1016/j.yebeh.2024.109825.

11 Vagus Nerve Stimulation (VNS)

11.1 Surgical Procedures

11.1.1 Anesthesia, Patient Positioning, and Skin Incision

The patient is placed in the supine position, and endotracheal intubation is performed using a Nerve Integrity Monitor (NIM®) electromyographic endotracheal tube (Medtronic, USA). This tube is an electromyographic monitoring endotracheal tube equipped with electrodes that monitor the activity of the vocal fold muscles. After the cervical vagus nerve is identified, electrical stimulation is applied with a stimulating probe, and the evoked responses are recorded in real time as electromyographic signals from the vocal folds. This technique ensures the accurate identification of the vagus nerve, distinguishing it from other cervical neural structures such as the ansa cervicalis. Anesthesia is performed using total intravenous anesthesia (TIVA), generally without the use of neuromuscular blocking agents.

The head is secured in the midline using a horseshoe headrest. The head is supported slightly below the level of the operating table, creating a vertex-down, chin-up position that provides cervical extension and facilitates the dissection of the sternocleidomastoid muscle (Fig. 11.1). A linear skin incision is made at the level of the cricoid cartilage, approximately 35 mm to the left of the midline. An additional straight incision of about 50 mm is made in the axillary region (Fig. 11.2). The length of the axillary incision depends on the selected generator model.

11.1.2 Identification/Exposure of the Vagus Nerve

After the cervical skin incision, we use a carotid endarterectomy retractor system to assist with exposure. This system enables four-directional retraction—superior, inferior, medial, and lateral—allowing the muscles to be retracted effectively and providing a well-expanded operative field (Fig. 11.2). The platysma is divided horizontally, exposing the sternocleidomastoid muscle, and dissection is carried out between the omohyoid and sternocleidomastoid. The lateral retraction of the sternocleidomastoid muscle exposes the internal jugular vein (Fig. 11.3). By dissecting along its medial aspect, the common carotid artery can be identified. At this stage, the lateral retraction of the internal jugular vein allows the identification of the vagus nerve, which usually runs between the common carotid artery and the internal jugular vein (Fig. 11.4). The nerve presumed to be the vagus is stimulated with the NIM monitoring system to confirm its identity. The vagus nerve is dissected free from the surrounding connective tissue. Placing a single 4–0 Nurolon

T. Uda et al., *Atlas of Epilepsy Surgery*, https://doi.org/10.1007/978-981-92-0176-1_11

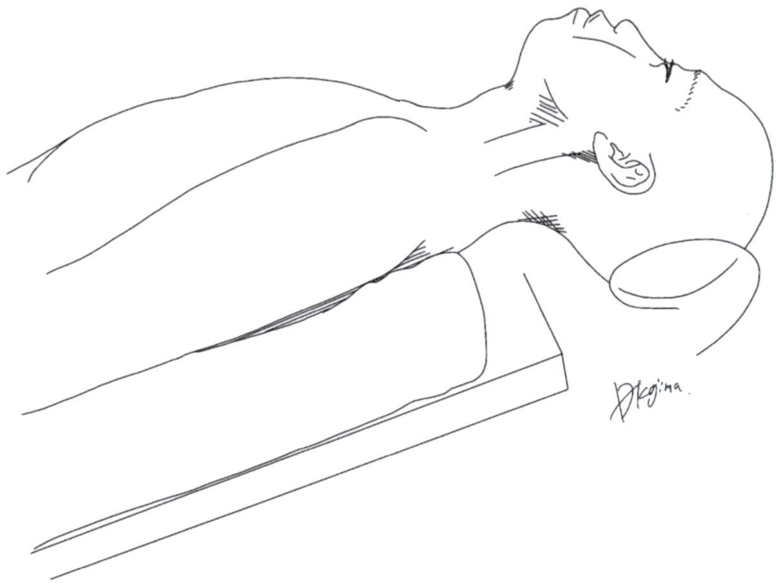

Fig. 11.1 Patient positioning for VNS. The head is secured in the midline using a horseshoe headrest. The head is supported slightly below the level of the operating table, creating a vertex-down, chin-up position that provides cervical extension and facilitates the dissection of the sternocleidomastoid muscle

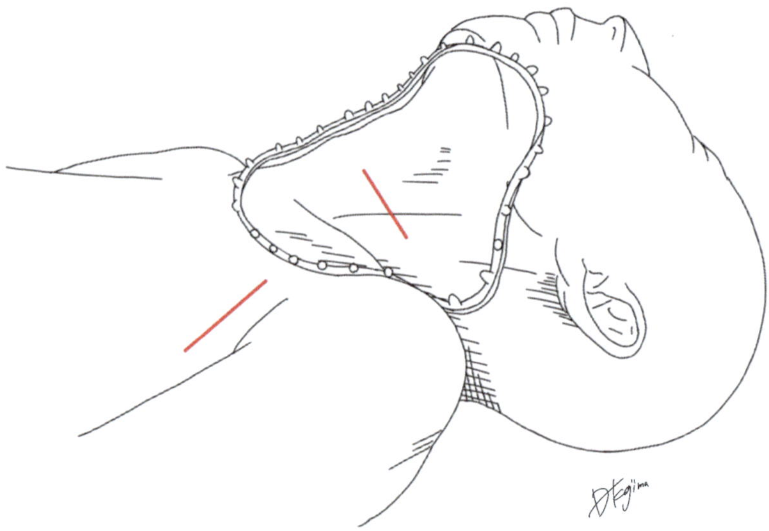

Fig. 11.2 Skin incision for VNS. A linear skin incision is made at the level of the cricoid cartilage, approximately 35 mm to the left of the midline. An additional straight incision of about 50 mm is made in the axillary region. A carotid endarterectomy retractor system was used to provide sufficient retraction for exposure

(Ethicon, USA) stitch on the nerve to apply gentle traction can facilitate the dissection. Exposure of at least 20 mm of the nerve generally provides adequate working space for wrapping the lead electrode around it. At this stage, a vessel loop is passed underneath the vagus nerve on both the cranial and caudal sides, and each is left loosely secured.

11.1.3 Subcutaneous Tunneling of the Lead and its Delivery to the Chest Pocket

To create the subcutaneous tunnel for the lead, the surgical field is temporarily shifted away from the cervical incision, and a skin incision is made in the axillary region, where a subcutane-

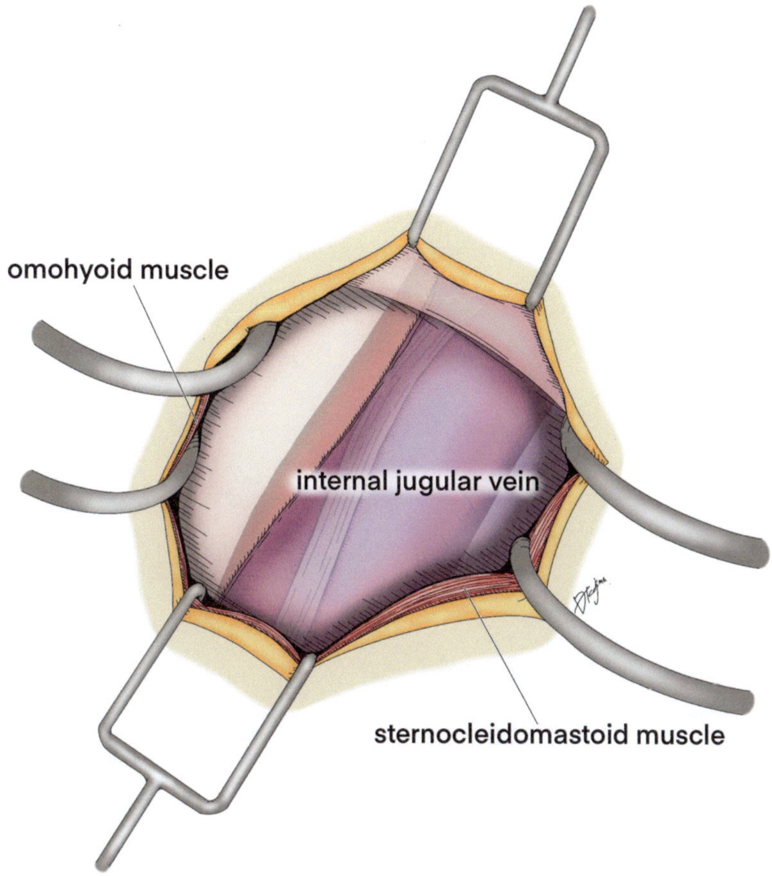

Fig. 11.3 Exposure of the vagus nerve. After dividing the platysma, dissection is carried out between the omohyoid and sternocleidomastoid muscles. Lateral retraction of the sternocleidomastoid muscle exposes the internal jugular vein

ous pocket is developed manually. The procedure then returns to the cervical surgical field, and the tunneler is introduced from the neck to create the subcutaneous tunnel, which is advanced until it reaches and communicates with the axillary subcutaneous pocket. The lead is then passed through the lumen of the tunneler and delivered from the cervical field to the axillary pocket.

11.1.4 Placement of the Lead Electrode on the Vagus Nerve

The exposed vagus nerve is reidentified after returning to the cervical surgical field. At this stage, we place a rubber sheet beneath the vagus nerve. Its primary purposes are to protect the surrounding structures, particularly the carotid sheath and the internal jugular vein, and to provide a clean background that facilitates visual distinction of the nerve during lead wrapping. The distal end of the lead has three helical coils, which are wrapped sequentially around the vagus nerve from the caudal coil to the cranial one (Fig. 11.5). After all three coils are wrapped, the vessel loops and the rubber sheet are removed from the field. The lead is then anchored between the sternocleidomastoid and omohyoid muscles using tie-downs.

11.1.5 Connection of the Lead to the Generator, Generator Implantation, and Closure

The proximal end of the lead, delivered to the axillary pocket, is connected to the generator.

Under the microscope, the connection is secured, and proper fixation is confirmed. After the setscrew is tightened with the torque wrench,

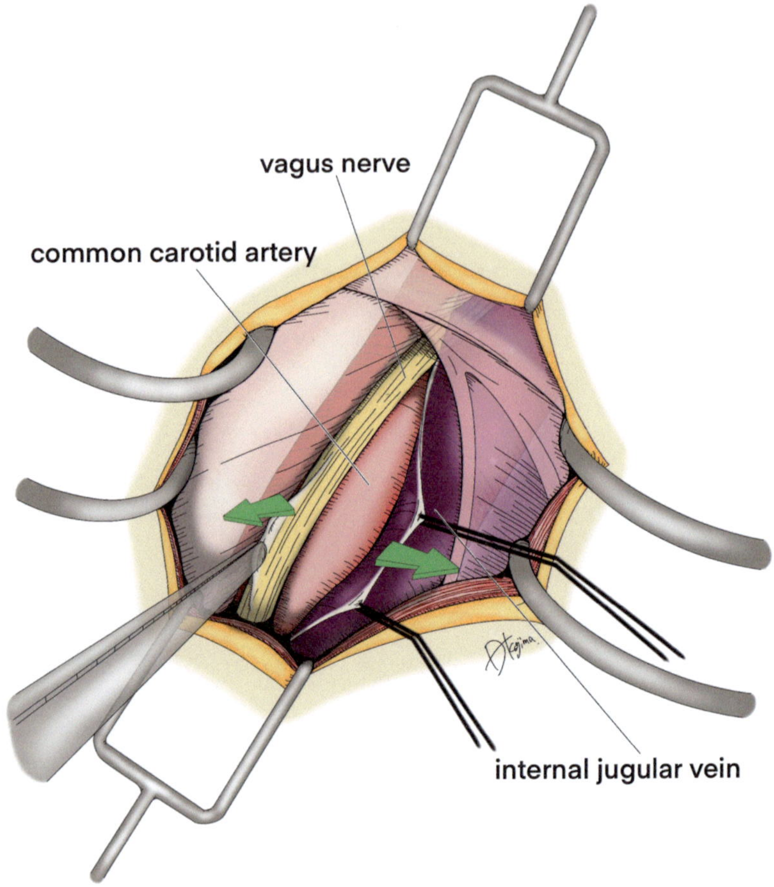

Fig. 11.4 Exposure of the vagus nerve. The lateral retraction of the internal jugular vein allows the identification of the vagus nerve, which usually runs between the common carotid artery and the internal jugular vein

system diagnostics and test stimulation are performed. Electrocardiogram (ECG) monitoring is used to confirm that bradycardia does not occur during stimulation. When using heart-rate-responsive generators such as AspireSR or Sentiva, it is recommended to confirm that the heart rate detected by the generator closely matches the heart rate displayed on the ECG monitor.

The generator is placed into the subcutaneous pocket and secured with a single absorbable suture. The lead is positioned deep, rather than superficial, to the generator to reduce the risk of inadvertent cutting during future generator replacement.

After the generator has been implanted and secured in the subcutaneous pocket, test stimulation is performed again to confirm that no abnormalities are detected. The cervical and axillary incisions are then closed, completing the procedure.

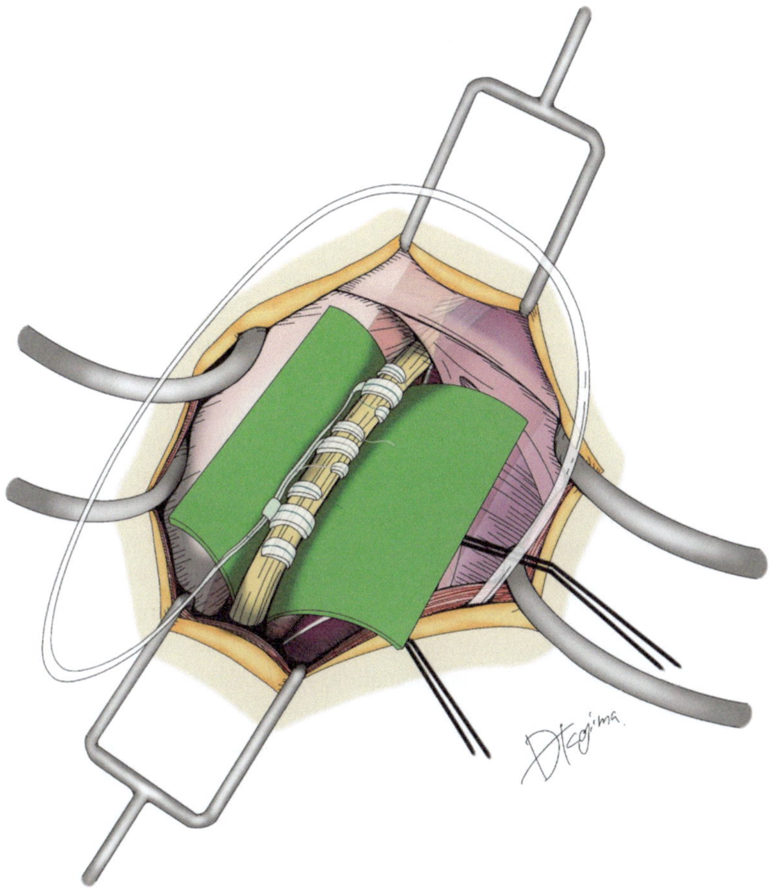

Fig. 11.5 Wrapping VNS electrodes. The distal end of the VNS lead consists of an anchor, a minus electrode, and a plus electrode. After placing a rubber sheet under the vagus nerve, the VNS electrode leads were wrapped sequentially

11.2 Notes

11.2.1 Procedure for VNS Generator Replacement

When the generator battery becomes depleted, a replacement procedure is required. During generator replacement, only the axillary region is operated on. A skin incision is made in the axilla, and the generator is removed. Care is taken to avoid injuring the lead during this step. The generator may sometimes be encapsulated by fibrous tissue. Once the generator has been safely removed, the subsequent steps follow the same procedure described in Sect. 11.1.5.

11.2.2 Methods for Identifying the Vagus Nerve

It is well known that the vagus nerve has variations in its running course in the cervical region. In a series of 326 cases evaluated with cervical ultrasonography, a retrocarotid course, in which the nerve runs posterior to the common carotid artery, was observed in 1.2% of patients [1]. In another report of 73 vagus nerve stimulation (VNS) cases, a retrocarotid course was observed in 4.1% [2]. In addition, pathological variations of the vagus nerve have been reported, including a case in a patient with neurofibromatosis type 1 presenting with a hypertrophied vagus nerve [3].

Given these findings, it is advisable to evaluate the course and diameter of the vagus nerve at the cervical incision level preoperatively. Ultrasonography can be used to evaluate the vagus nerve; however, this requires operator familiarity with interpreting the nerve on ultrasound. To address this, we have reported a preoperative method for visualizing the vagus nerve using fusion imaging of cervical magnetic resonance imaging (MRI) and three-dimensional (3D) cervical computed tomography angiography (CTA) [4]. We have since refined this technique and now visualize the vagus nerve using 3D CTA alone.

In addition to the vagus nerve, other cervical nerves, such as the ansa cervicalis, may be encountered during VNS surgery, and these structures may be mistaken for the vagus nerve. To ensure accurate identification, we use NIM, which detects pharyngeal and laryngeal muscle responses evoked by electrical stimulation. This physiological feedback confirms that the stimulated structure is indeed the vagus nerve. Pediatric versions of the tube (inner diameter 5.0 mm) are now available and can be used in children aged 4 years and older. With these preoperative visualization techniques and intraoperative monitoring, the vagus nerve can be reliably identified in most cases. However, cases have been reported in which the superior root of the ansa cervicalis joins the vagus nerve, and therefore, careful attention is required during electrode placement [5].

11.2.3 Efficacy of VNS for Epileptic Seizures

Early double-blind, randomized controlled trials (E03 and E05) demonstrated that patients receiving adjunctive VNS therapy experienced an approximately 25–30% reduction in seizure frequency from baseline, indicating a statistically significant therapeutic effect [6, 7]. Long-term follow-up studies and large registry analyses have consistently shown that the therapeutic effect of VNS increases over time, demonstrating a progressive (time-dependent) improvement in seizure control. Englot et al. analyzed 5554 cases from the VNS Patient Outcome Registry and reported that the responder rate, defined as a $\geq$ 50% reduction in seizure frequency, reached 49% within 4 months and 63% within 48 months. In their systematic literature review of 2869 patients across 78 studies, the corresponding responder rates were 40.0% at 4 months and 60.1% at 48 months [8]. Morris et al. evaluated data from five studies (E01–E05) conducted between 1988 and 1997 and demonstrated progressively improving outcomes, with responder rates of 36.8% at 1 year, 43.2% at 2 years, and 42.7% at 3 years [9]. Similarly, Elliott et al. reported that among their cohort of 65 patients, including 21 children, the responder rates were 35.7% at 6 months, 52.1% at 1 year, 58.3% at 2 years, and exceeded 60% after 4 years [10], confirming that the progressive enhancement of VNS efficacy occurs across age groups. The American Academy of Neurology (AAN) updated its guidelines in 2013, classifying the efficacy of VNS as a Level C recommendation based on Class I–II evidence [11]. The guidelines also state that "VNS may be considered progressively effective in patients over multiple years of exposure (Level C)." Taken together, these findings indicate that VNS is considered a therapy whose seizure-reducing effects progressively strengthen over the long term.

11.2.4 Effectiveness of VNS for Other Disorders

VNS has also been explored for a variety of neurological and psychiatric disorders beyond epilepsy. Among these, treatment-resistant depression (TRD) is the condition for which the most consistent long-term benefits have been reported. Aaronson et al. demonstrated in a 5-year registry study that patients receiving adjunctive VNS achieved a cumulative response rate of 67.6% and a cumulative remission rate of 43.3% [12].

In the RECOVER trial, all patients received VNS implantation, and a prospective randomized comparison was conducted between stimulation-

ON and stimulation-OFF groups [13]. Although no significant difference in remission rate was observed at 12 months—the primary endpoint—the between-group difference was most pronounced during the final 10–12 months of the study period. In this study, both groups showed clinical improvement. However, the stimulation-ON group demonstrated a clearer trend toward long-term benefit, indicating that prolonged stimulation may provide increasing therapeutic effects.

In addition, a randomized controlled trial involving 493 adults with treatment-resistant depression demonstrated significant therapeutic effects of VNS across multiple depression rating scales, and VNS was reported to be both safe and effective in this population [14].

Thus, accumulating evidence suggests that VNS may exert beneficial effects in depression through the modulation of neural networks, and further long-term STUDIES are anticipated to clarify its therapeutic potential.

References

1. Park JK, Jeong SY, Lee JH, Lim GC, Chang JW. Variations in the course of the cervical vagus nerve on thyroid ultrasonography. AJNR Am J Neuroradiol. 2011;32(7):1178–81. https://doi.org/10.3174/ajnr.A2476.
2. Takamizawa S, Morino M. Knowledge of anatomical relationships of the vagus nerve with internal jugular vein and common carotid artery in the neck necessary for vagus nerve stimulation (VNS) surgery. Jpn J Neurosurg (Tokyo). 2015;24(9):705–11. In Japanese
3. Hamdi H, Brun G, Zanello M, Dibué M, Beltaifa Y, Lagarde S, et al. VNS implantation in a NF1 patient: massive nerve hypertrophy discovered intra-operatively preventing successful electrode placement. Case report. Acta Neurochir (Wien). 2020;162(10):2509–12. https://doi.org/10.1007/s00701-020-04535-y.
4. Nakae S, Kumon M, Katagata A, Murayama K, Hirose Y. Vagus nerve visualization using fused images of 3D-CT angiography and MRI as preoperative evaluation for vagus nerve stimulation. Brain Sci. 2023;13(3):396. https://doi.org/10.3390/brainsci13030396.
5. Gopalakrishnan CV, Kestle JRW, Connolly MB. The “vagal ansa”: a source of complication in vagus nerve stimulation. J Neurosurg Pediatr. 2015;15(5):535–8. https://doi.org/10.3171/2014.10.PEDS14259.
6. Handforth A, DeGiorgio CM, Schachter SC, Uthman BM, Naritoku DK, Tecoma ES, et al. Vagus nerve stimulation therapy for partial-onset seizures: a randomized active-control trial. Neurology. 1998;51(1):48–55. https://doi.org/10.1212/WNL.51.1.48.
7. Vagus Nerve Stimulation Study Group. A randomized controlled trial of chronic vagus nerve stimulation for treatment of medically intractable seizures. Neurology. 1995;45(2):224–30. https://doi.org/10.1212/WNL.45.2.224.
8. Englot DJ, Rolston JD, Wright CW, Hassnain KH, Chang EF. Rates and predictors of seizure freedom with vagus nerve stimulation for intractable epilepsy. Neurosurgery. 2016;79(3):345–53. https://doi.org/10.1227/NEU.0000000000001165.
9. Morris GL 3rd, Mueller WM. Long-term treatment with vagus nerve stimulation in patients with refractory epilepsy. Vagus nerve stimulation study group E01–E05. Neurology. 1999;53(8):1731–5. https://doi.org/10.1212/WNL.53.8.1731.
10. Elliott RE, Morsi A, Tanweer O, Grobelny B, Geller E, Carlson C, et al. Efficacy of vagus nerve stimulation over time: review of 65 consecutive patients with treatment-resistant epilepsy treated with VNS >10 years. Epilepsy Behav. 2011;20(3):478–83. https://doi.org/10.1016/j.yebeh.2010.12.042.
11. Morris GL 3rd, Gloss D, Buchhalter J, Mack KJ, Nickels K, Harden C, et al. Evidence-based guideline update: vagus nerve stimulation for the treatment of epilepsy: report of the guideline development Subcommittee of the American Academy of neurology. Neurology 2013;81(16):1453–1459. doi:https://doi.org/10.1212/WNL.0b013e3182a393d1.
12. Aaronson ST, Sears P, Ruvuna F, Bunker M, Conway CR, Dougherty DD, et al. A 5-year observational study of patients with treatment-resistant depression treated with vagus nerve stimulation or treatment as usual: comparison of response, remission, and suicidality. Am J Psychiatry. 2017;174(7):640–8. https://doi.org/10.1176/appi.ajp.2017.16010034.
13. Sackeim HA, Conway CR, Aaronson ST, Bunker MT, Gordon C, Y-CL L, et al. Characterizing the effects of vagus nerve stimulation on symptom improvement in markedly treatment-resistant major depressive disorder: a RECOVER trial report. J Affect Disord. 2025;380:135–45. https://doi.org/10.1016/j.jad.2025.03.124.
14. Conway CR, Aaronson ST, Sackeim HA, George MS, Zajecka J, Bunker MT, et al. Vagus nerve stimulation in treatment-resistant depression: a one-year, randomized, sham-controlled trial. Brain Stimul. 2025;18(3):676–89. https://doi.org/10.1016/j.brs.2024.12.1191.

Future Directions in Focal Epilepsy Surgery

12

12.1 Ablation

12.1.1 Radiofrequency Thermocoagulation (RF-TC)

Radiofrequency thermocoagulation (RF-TC) is a minimally invasive ablative treatment for patients with focal drug-resistant epilepsy. This technique delivers high-frequency electrical current via stereotactically implanted depth electrodes, producing localized thermal coagulation to ablate epileptogenic lesions or seizure onset zones (SOZs). A major advantage of RF-TC is that, when combined with stereoelectroencephalography (SEEG) for seizure localization, it enables a seamless transition from diagnostic electrode implantation to therapeutic intervention. Because RF-TC can be performed using implanted SEEG electrodes without the need for additional dedicated devices, it has gained widespread acceptance as a practical and minimally invasive treatment option. However, as the ablation effect is confined to the immediate vicinity of the electrode contacts, a careful assessment of the spatial extent of the epileptogenic lesion or focus and appropriate target selection are essential.

RF-TC is particularly useful for patients with well-localized focal epilepsy or deep-seated lesions that are difficult to resect surgically. In selected cases, significant seizure reduction or even seizure freedom has been reported. Previous studies have demonstrated that SEEG-guided RF-TC results in a $\geq$ 50% reduction in seizure frequency in approximately half of treated patients at 1-year follow-up, with a low risk of complications [1–3]. Higher rates of seizure freedom have been reported in specific pathological conditions, such as nodular heterotopia, hypothalamic hamartoma, and focal cortical dysplasia [3]. Kreinter et al. identified ablation of the SOZ as the most important predictor of treatment outcome [4].

In contrast, regions involved in early seizure propagation are also likely to constitute components of the epileptogenic network, and their ablation may contribute to seizure reduction. Indeed, the ablation of all contacts involved in the SOZ has been associated with higher seizure freedom and responder rates, whereas the ablation of early propagation areas appears to improve the responder rate but does not necessarily translate into higher seizure freedom rates [4].

These findings suggest that RF-TC is not only a low-risk, minimally invasive therapeutic option but may also serve as a functional test for epileptogenicity. The evaluation of ablation targets and the resulting responder rate may provide valuable prognostic information for predicting the efficacy of subsequent definitive epilepsy surgery.

More recently, in addition to RF-TC performed through SEEG electrodes, volume-based RF-TC using dedicated coagulation probes has been reported [5, 6]. Unlike point-based ablations between electrode contacts, this approach allows the creation of a three-dimensional and more homogeneous thermal lesion around the

T. Uda et al., *Atlas of Epilepsy Surgery*, https://doi.org/10.1007/978-981-92-0176-1_12

probe tip, resulting in a larger and more uniform coagulation volume. This technique may be particularly advantageous in cases involving deep-seated structures, such as the insula, or in patients with more extensive epileptogenic regions. Volume-based RF-TC therefore represents a promising advancement in the evolving field of minimally invasive focal epilepsy surgery.

12.1.2 Other Ablative Techniques

Laser interstitial thermal therapy (LITT) is an ablative treatment in which a laser fiber is stereotactically inserted into the target lesion, and thermal energy generated by the laser is used to induce coagulative necrosis [7]. The procedure is performed in a magnetic resonance imaging (MRI) environment, allowing the real-time monitoring of temperature changes using MR thermometry. This enables high spatial accuracy in targeting the intended lesion [8, 9].

LITT has primarily been reported in the treatment of mesial temporal sclerosis, hypothalamic hamartoma, and relatively small focal lesions located in deep brain structures and is regarded as one of the minimally invasive treatment options [8]. However, the requirement for specialized equipment, the associated high procedural costs, and the limited availability of appropriately equipped centers have restricted its clinical application largely to specific regions and institutions, particularly in North America.

Focused ultrasound (FUS) is an ablative technique that delivers convergent ultrasound energy from outside the skull to heat a target region and induce coagulative necrosis [10]. Similar to LITT, FUS is performed in an MRI environment for both target localization and the monitoring of temperature changes during treatment.

The principal advantage of FUS lies in its completely noninvasive nature, as it does not require cranial penetration or craniotomy. Two distinct therapeutic concepts exist within FUS-based interventions: ablative treatment using high-intensity focused ultrasound (HIFU) and neuromodulation using low-intensity focused ultrasound (LIFU) [11]. LIFU has been investigated as a novel neuromodulatory approach that reversibly alters neuronal activity and may suppress hyperexcitability within epileptic networks. However, FUS is strongly affected by ultrasound attenuation and distortion caused by the skull, and there is also a risk of skull overheating, which imposes substantial limitations on focal accuracy and energy delivery [11]. In addition, the location and size of lesions amenable to treatment are restricted, and the clinical application of FUS in epilepsy has so far remained largely at the level of case reports and early-stage research.

Taken together, although FUS holds considerable potential as a noninvasive treatment modality for drug-resistant epilepsy, it has not yet reached a stage at which it can replace established surgical approaches or neurostimulation therapies. Further studies are required to clarify appropriate patient selection, optimize sonication parameters, and evaluate long-term outcomes. At present, FUS is therefore best regarded not as a practical alternative to RF-TC or LITT but rather as an emerging technique that illustrates the future possibilities of noninvasive ablative treatment.

12.1.3 Neuromodulation Other than VNS

12.1.4 Deep Brain Stimulation (DBS)

Deep brain stimulation (DBS) was approved as another neuromodulation therapy for epilepsy. DBS is currently indicated for the stimulation of the anterior nucleus of the thalamus in patients with focal-onset seizures. A pivotal trial underlying this indication is the SANTE (Stimulation of the Anterior Nucleus of the Thalamus for Epilepsy) study. The SANTE trial was a multicenter, prospective, randomized, double-blind study that evaluated the efficacy and safety of DBS targeting the anterior nucleus of the thalamus in patients with drug-resistant focal epilepsy. In the initial report, median seizure frequency was reduced by 56% at 2 years, with a responder rate of 54% [12]. At 5 years, a 69% reduction in seizure frequency was reported, and the responder rate reached 68% [13]. Another well-known investigation is the MORE (Medtronic Registry for Epilepsy) study. This multicenter, prospective registry investigated the efficacy and safety of DBS, targeting the anterior nucleus of the thalamus in patients with drug-resistant focal epilepsy. Unlike the SANTE trial, the MORE study evaluated real-world outcomes by collecting data from patients who underwent anterior thalamic DBS [14]. In 2024, its long-term outcomes were reported, demonstrating that the efficacy of anterior thalamic DBS continued to improve over time. Median seizure frequency was reduced by 56% at 5 years, and the responder rate at the final follow-up reached 41%. These findings indicate that anterior thalamic DBS remains safe and effective during long-term use for up to 5 years [15].

Studies comparing DBS and VNS have also been reported. In a meta-analysis by Skrehot et al., DBS and responsive neurostimulation (RNS) showed significantly greater seizure-reduction efficacy at 1 year than VNS in patients with drug-resistant focal epilepsy [16]. Although many reports suggest that DBS may provide greater seizure suppression than VNS [17], no randomized controlled trials have directly compared the two modalities. Moreover, because there are reports that DBS can worsen depressive symptoms, VNS may be chosen in cases with comorbid psychiatric conditions or depending on the clinical context.

In addition, studies of DBS targeting generalized seizures have been actively conducted. The centromedian nucleus (CMN) of the thalamus and the pulvinar nucleus have been reported as the principal stimulation targets [18, 19]. These targets have shown potential efficacy primarily in generalized epilepsy, and further accumulation of clinical evidence is anticipated.

12.1.5 Responsive Neurostimulation (RNS)

Responsive neurostimulation (RNS) is a closed-loop neuromodulation therapy that continuously records intracranial electrocorticography using implanted electrodes and delivers electrical stimulation in response to the detection of seizure-associated abnormal brain activity [20]. In this context, closed loop refers to a therapeutic paradigm in which cortical electrophysiological activity is monitored in real time and stimulation is triggered in response to predefined seizure-related electroencephalographic patterns [21].

In RNS therapy, recorded electrocorticographic data are used to evaluate stimulation responses, and device programming is individualized for each patient to optimize detection algorithms and stimulation parameters in order to maximize seizure suppression. A key characteristic of RNS is that stimulation is delivered only during seizure or the preictal period, in contrast to other neuromodulation therapies that provide continuous or cyclic stimulation.

The RNS system allows simultaneous recording and stimulation from up to two seizure onset zones or seizure-related network nodes using intracranial electrodes [21]. This capability makes RNS a potential therapeutic option for patients with focal epilepsy that is not amenable to resection, as well as for selected cases of multifocal epilepsy.

The efficacy of RNS in adults with drug-resistant epilepsy has been supported by Class I evidence reported by Morrell et al. [21] At the end of a 12-week double-blind evaluation period, patients receiving active stimulation demonstrated a significantly greater reduction in seizure frequency compared with those receiving sham stimulation (37.9% vs. 17.3%). Moreover, treatment efficacy was shown to increase over time, with mean seizure reductions of 44% at 1 year and 53% at 2 years of follow-up.

The application of RNS has also been explored in pediatric patients and in patients with generalized epilepsy. In a series of 27 pediatric patients reported by Panov et al., 15 of 22 patients (68.2%) achieved a $\geq$ 50% reduction in seizure frequency at 1 year postoperatively [22]. In addition, long-term seizure reduction has been suggested in patients with Lennox-Gastaut syndrome, multi-focal epilepsy, and generalized seizures [23].

Regarding safety, Morrell et al. reported that serious intracranial hemorrhage without permanent neurological deficit occurred in 2.1% of patients, and serious infection-related adverse events occurred in 4.2% [21]. Importantly, responsive cortical stimulation with RNS was not associated with a significant increase in the risk of depression or suicidal ideation [21].

Taken together, RNS represents a neuromodulation therapy that directly monitors and modulates cortical activity in the vicinity of the epileptogenic focus, and it is characterized by therapeutic properties distinct from those of continuous stimulation therapies such as DBS and VNS. Accordingly, optimal treatment selection requires a careful consideration of the individual patient's clinical characteristics, seizure network organization, and therapeutic goals.

12.2 Endoscopic and Exoscopic Surgeries in Focal Epilepsy

12.2.1 Concept and Technical Considerations

Neuroendoscopic and exoscopic techniques are now widely used across various fields of neurosurgical practice, and their application in epilepsy surgery has also increased steadily over recent years. In particular, for the evaluation of focal epilepsy using intracranial electroencephalography, SEEG has increasingly become the dominant approach. One of the major reasons for this shift is its minimally invasive nature. If SEEG is selected primarily to minimize invasiveness, it is reasonable to consider that subsequent focal resection should also aim to maintain a similarly low degree of invasiveness. Against this background, we have actively introduced both endoscopic and exoscopic techniques into the focal resection of epilepsy. Although exoscopic surgery is comparable to conventional microscopic surgery in terms of invasiveness, it shares the head-up surgical posture characteristic of endoscopic surgery. From this perspective, we regard exoscopic surgery as an important transitional modality that bridges conventional microscopic techniques and fully endoscopic procedures.

Endoscopic surgery offers several advantages, including the ability to minimize the extent of craniotomy and to obtain a wide operative field in deep anatomical regions with fewer blind spots, even within a narrow surgical corridor. In contrast, hemostatic manipulation may be more challenging compared with conventional microscopic surgery, which provides a wider working space and allows stable bimanual operation. Consequently, endoscopic surgery may not always be optimal for highly vascular lesions or in situations associated with an increased risk of bleeding. Therefore, when performing epilepsy focal resection using neuroendoscopic techniques, it is essential not to prioritize minimal invasiveness alone but rather to select the most appropriate surgical approach based on anatomical considerations and the pathological characteristics of the lesion.

12.2.2 Preferred Cases for Endoscopic and Exoscopic Approaches

The selection of endoscopic or exoscopic approaches is based on the location and characteristics of the lesion.

Specifically, cases involving small lesions, such as cerebral cavernous malformations (CCMs), located in the subcallosal area—including the gyrus rectus and orbital gyri—are considered good candidates for endoscopic surgery. When lesions in this region are resected using conventional microscopic techniques, a bicoronal skin incision is generally required. In contrast, endoscopic surgery can be performed through a linear skin incision, allowing a substantial reduction in both the extent of the skin incision and the craniotomy. In addition, regardless of whether SEEG has been performed beforehand, the preoperative placement of depth electrodes in the target region as navigation guides not only enables intraoperative electroencephalogram (EEG) monitoring but also facilitates orientation during surgery, thereby reducing the risk of intraoperative disorientation. This strategy represents a useful option in endoscopic epilepsy surgery.

In approaches that access lesions via the posterior cranial fossa, such as the supracerebellar transtentorial (SCTT) approach, the use of a conventional operating microscope may limit the angle of view because of the relatively large size of the device and the restricted freedom of the supporting arm. As a result, when targeting deep-seated lesions located in the posterior medial temporal lobe—such as the posterior parahippocampal gyrus or the lingual gyrus—or in the medial occipital lobe, it can be difficult to maintain adequate focus on the lesion. This limitation often necessitates a larger craniotomy.

Accordingly, when selecting the SCTT approach, the use of an exoscope, an endoscope, or a combination of both may increase the flexibility of viewing angles and thereby enable a less invasive surgical procedure.

Beyond these applications, various endoscopic techniques for epilepsy focal resection have been reported, including endoscopic hemispherotomy [24], endoscopic hippocampectomy [25], and intraventricular endoscopic approaches for hypothalamic hamartomas [7]. These techniques have been refined according to institutional experience and patient-specific factors. With the continued accumulation of clinical experience, the indications for and the roles of endoscopic and exoscopic techniques in epilepsy focal resection are expected to expand further in the future.

12.3 Network-Based Concepts in Focal Epilepsy Surgery

In recent years, epilepsy surgery has increasingly focused not only on the SOZ but also on the broader epileptogenic network, which includes cortical regions involved in early seizure spread. Although these regions are not necessarily responsible for seizure initiation, they may contribute to seizure maintenance and the generation of clinical symptoms and, in some cases, may influence surgical outcomes.

A systematic review examining the relationship between seizure propagation patterns and postoperative seizure outcomes suggested that early seizure spread may reflect the underlying seizure network. In particular, cortical regions involved in seizure propagation within 3–10 seconds after seizure onset were found to be associated with postoperative seizure freedom. It has also been pointed out that the preservation of cortical areas involved in early seizure spread may increase the risk of postoperative seizure recurrence [26].

More recently, the concept of in silico surgery has been proposed, in which seizure dynamics following hypothetical surgical interventions are predicted using computational simulations based on SEEG data. Several studies, primarily led by the group of Bartolomei et al., have demonstrated the feasibility of virtual surgical approaches using patient-specific brain network models [27, 28]. These virtual brain models integrate individual MRI data with machine learning techniques and optimization algorithms to estimate the extent and composition of the epileptogenic zone (EZ), as well as the spatiotemporal dynamics of seizure initiation and propagation. The primary objective of this framework is to reconstruct patient-specific seizure networks [23, 24]. Research aimed at developing network-based therapeutic strategies is currently underway.

Although this approach remains at the research stage, the combination of SEEG-based network analysis with minimally invasive treatment modalities may contribute to the future personalization and optimization of epilepsy focal resection.

References

1. Guénot M, Isnard J, Ryvlin P, Fischer C, Mauguière F, Sindou M. SEEG-guided RF thermocoagulation of epileptic foci: feasibility, safety, and preliminary results. Epilepsia. 2004;45(11):1368–74. https://doi.org/10.1111/j.0013-9580.2004.17704.x.

2. Bourdillon P, Isnard J, Catenoix H, Montavont A, Rheims S, Ryvlin P, et al. Stereo electroencephalography-guided radiofrequency thermocoagulation (SEEG-guided RF-TC) in drug-resistant focal epilepsy: results from a 10-year experience. Epilepsia. 2017;58(1):85–93. https://doi.org/10.1111/epi.13616.
3. Bourdillon P, Rheims S, Catenoix H, Montavont A, Ostrowsky-Coste K, Isnard J, et al. Surgical techniques: stereoelectroencephalography-guided radiofrequency thermocoagulation (SEEG-guided RF-TC). Seizure. 2020;77:64–8. https://doi.org/10.1016/j.seizure.2019.01.021.
4. Kreinter H, Espino PH, Mejía S, Alorabi K, Gilmore G, Burneo JG, et al. Disrupting the epileptogenic network with stereoelectroencephalography-guided radiofrequency thermocoagulation. Epilepsia. 2024;65(7):e113–8. https://doi.org/10.1111/epi.18005.
5. Takayama Y, Kimura Y, Iijima K, Yokosako S, Kosugi K, Yamamoto K, et al. Volume-based radiofrequency thermocoagulation for pediatric insulo-opercular epilepsy: a feasibility study. Oper Neurosurg (Hagerstown). 2022;23(3):241–9. https://doi.org/10.1227/ons.0000000000000294.
6. Mullatti N, Landre E, Mellerio C, Oliveira AJ, Laurent A, Turak B, et al. Stereotactic thermocoagulation for insular epilepsy: lessons from successes and failures. Epilepsia. 2019;60(8):1565–79. https://doi.org/10.1111/epi.16092.
7. Bown SG. Phototherapy in tumors. World J Surg. 1983;7:700–9. https://doi.org/10.1007/BF01655209.
8. Infante N, Conesa G, Pérez-Enríquez C, Capellades J, Panadés de Oliveira L, Vilella L, et al. MRI-guided laser interstitial thermal therapy in epilepsy: indications, technique and outcome in an adult population. A single-center data analysis. Acta Neurochir. 2025;167(1):39 https://doi.org/10.1007/s00701-025-06429-3.
9. Cossu G, Sufianov R, Messerer M, Ryvlin P, Daniel RT, Sufianov A. Intraventricular endoscopic surgery for pediatric hypothalamic hamartomas: surgical series and literature meta-analysis. World Neurosurg. 2025;204:124505. https://doi.org/10.1016/j.wneu.2025.124505.
10. Quadri SA, Waqas M, Khan I, Khan MA, Suriya SS, Farooqui M, et al. High-intensity focused ultrasound: past, present, and future in neurosurgery. Neurosurg Focus. 2018;44(2):E16. https://doi.org/10.3171/2017.11.FOCUS17610.
11. Lescrauwaet E, Vonck K, Sprengers M, Raedt R, Klooster D, Carrette E, et al. Recent advances in the use of focused ultrasound as a treatment for epilepsy. Front Neurosci. 2022;16:886584. https://doi.org/10.3389/fnins.2022.886584.
12. Fisher R, Salanova V, Witt T, Worth R, Henry T, Gross R, et al. Electrical stimulation of the anterior nucleus of thalamus for treatment of refractory epilepsy. Epilepsia. 2010;51(5):899–908. https://doi.org/10.1111/j.1528-1167.2010.02536.x.
13. Salanova V, Witt T, Worth R, Henry TR, Gross RE, Nazzaro JM, et al. Long-term efficacy and safety of thalamic stimulation for drug-resistant partial epilepsy. Neurology. 2015;84(10):1017–25. https://doi.org/10.1212/WNL.0000000000001334.
14. Peltola J, Colon AJ, Pimentel J, Coenen VA, Gil-Nagel A, Gonçalves Ferreira A, et al. Deep brain stimulation of the anterior nucleus of the thalamus in drug-resistant epilepsy in the MORE multicenter patient registry. Neurology. 2023;100(18):e1852–65. https://doi.org/10.1212/WNL.0000000000206887.
15. Kaufmann E, Peltola J, Colon AJ, Lehtimäki K, Majtanik M, Mai JK, et al. Long-term evaluation of anterior thalamic deep brain stimulation for epilepsy in the European MORE registry. Epilepsia. 2024;65(8):2438–58. https://doi.org/10.1111/epi.18003.
16. Skrehot HC, Englot DJ, Haneef Z. Neuro-stimulation in focal epilepsy: a systematic review and meta-analysis. Epilepsy Behav. 2023;142:109182. https://doi.org/10.1016/j.yebeh.2023.109182.
17. Zhu J, Wang X, Xu C, Zhang X, Qiao L, Zhang X, et al. Comparison of efficiency between VNS and ANT-DBS therapy in drug-resistant epilepsy: a one year follow up study. J Clin Neurosci. 2021;90:112–7. https://doi.org/10.1016/j.jocn.2021.05.046.
18. Valentín A, García Navarrete E, Chelvarajah R, Torres C, Navas M, Vico L, et al. Deep brain stimulation of the centromedian thalamic nucleus for the treatment of generalized and frontal epilepsies. Epilepsia. 2013;54(10):1823–33. https://doi.org/10.1111/epi.12352.
19. Chandran AS, Joshi S, Suresh S, Savarraj J, Snyder K, Vasconcellos FN, et al. Efficacy of neuromodulation of the pulvinar nucleus for drug-resistant epilepsy. Epilepsia. 2025;66(4):1059–70. https://doi.org/10.1111/epi.18244.
20. Venetucci Gouveia F, Warsi NM, Suresh H, Matin R, Ibrahim GM. Neurostimulation treatments for epilepsy: deep brain stimulation, responsive neurostimulation and vagus nerve stimulation. Neurotherapeutics. 2024;21(3):e00308. https://doi.org/10.1016/j.neurot.2023.e00308.
21. Morrell MJ; RNS System in Epilepsy Study Group. Responsive cortical stimulation for the treatment of medically intractable partial epilepsy: a randomized controlled trial. Neurology 2011;77(13):1295–1304. doi:https://doi.org/10.1212/WNL.0b013e3182302056.
22. Panov F, Ganaha S, Haskell J, Fields M, La Vega-Talbott M, Wolf S, et al. Safety of responsive neurostimulation in pediatric patients with medically refractory epilepsy. J Neurosurg Pediatr. 2020;26(5):525–32. https://doi.org/10.3171/2020.5.PEDS20118.
23. Roa JA, Abramova M, Fields M, La Vega-Talbott M, Yoo J, Marcuse L, et al. Responsive neurostimulation of the thalamus for the treatment of refractory epilepsy. Front Hum Neurosci 2022;16:926337. doi:https://doi.org/10.3389/fnhum.2022.926337.

24. Koh S, Uda T, Kunihiro N, Umaba R, Ishimoto K, Kawashima T, et al. Endoscopic parasagittal vertical hemispherotomy: a 2-dimensional operative video. Oper Neurosurg (Hagerstown). 2025;29(4):598–9. https://doi.org/10.1227/ons.0000000000001466.
25. Uda T, Kawashima T, Hattori M, Kojima Y, Mito Y, Goto T. Endoscopic transcortical selective amygdalohippocampectomy for mesial temporal lobe epilepsy: 2-dimensional operative video. Oper Neurosurg (Hagerstown). 2021;21(5):E443. https://doi.org/10.1093/ons/opab265.
26. Andrews JP, Ammanuel S, Kleen J, Khambhati AN, Knowlton R, Chang EF. Early seizure spread and epilepsy surgery: a systematic review. Epilepsia. 2020;61(10):2163–72. https://doi.org/10.1111/epi.16668.
27. Dollomaja B, Wang HE, Guye M, Makhalova J, Bartolomei F, Jirsa VK. Virtual epilepsy patient cohort: generation and evaluation. PLoS Comput Biol. 2025;21(4):e1012911. https://doi.org/10.1371/journal.pcbi.1012911.
28. Jirsa V, Wang H, Triebkorn P, Hashemi M, Jha J, Gonzalez-Martinez J, Guye M, Makhalova J, Bartolomei F. Personalised virtual brain models in epilepsy. Lancet Neurol. 2023;22(5):443–54. https://doi.org/10.1016/S1474-4422(23)00008-X.

Index

T. Uda et al., *Atlas of Epilepsy Surgery*, https://doi.org/10.1007/978-981-92-0176-1

MIX
Papier aus verantwortungsvollen Quellen
Paper from responsible sources
FSC® C105338

If you have any concerns about our products,
you can contact us on
ProductSafety@springernature.com

In case Publisher is established outside the EU,
the EU authorized representative is:
Springer Nature Customer Service Center GmbH
Europaplatz 3, 69115 Heidelberg, Germany

Printed by Libri Plureos GmbH
in Hamburg, Germany